未来之境
扎哈·哈迪德建筑事务所设计展

THE NEW WORLD
Zaha Hadid Architects

扎哈·哈迪德建筑事务所　嘉德艺术中心　编
Zaha Hadid Architects　Guardian Art Center

广西师范大学出版社
·桂林·

从"心"定义未来

寇勤
嘉德艺术中心总经理

扎哈·哈迪德是建筑设计领域的标杆。她是第一位赢得普利兹克奖项的女性建筑设计师，同时为中国建筑设计领域做出了杰出贡献。2010年，广州大剧院竣工之后，扎哈·哈迪德建筑事务所在中国已经有北京丽泽SOHO和大兴国际机场等14个已建成的项目以及25个正在建设的项目，为我们的城市增添了活力与激情。

一直以来，嘉德艺术中心致力于将海内外高质量的艺术展览奉献给观众。此次"未来之境——扎哈·哈迪德建筑事务所设计展"在嘉德艺术中心揭幕，将事务所建成以来四十年间在世界各地的作品一并呈现给我们的观众，讲述了事务所如何通过新的数字设计工具、机器人技术、3D打印技术、人工智能、元宇宙虚拟建筑等不断追求、探索建筑设计中的创新，同时展示了事务所对于建筑行业可持续性的关注，以及其在环保领域广受赞誉的项目研究成果。在北京举办一场最为盛大的建筑设计展，是扎哈·哈迪德女士和其事务所长久以来的愿望，如今终于与嘉德艺术中心携手实现。借此机会，嘉德文库联手扎哈·哈迪德建筑事务所共同推出展览同名图书，定格此次建筑艺术展的精彩瞬间，并作为展览内容的补充与延续。在新世界的语境中，扎哈·哈迪德建筑事务所回应当下挑战，展望未来，表达了对未来坚定不移的乐观态度。希望观众能够通过这本《未来之境：扎哈·哈迪德建筑事务所设计展》，自由穿越于扎哈·哈迪德建筑事务所构建的城市片段之中，重新聚焦人与自然，思考所谓"存在"，获得共鸣，汲取力量，从"心"定义未来。

Redefining the Future in a Heartfelt Way

Kou Qin
General Manager of Guardian Art Center

Zaha Hadid, an icon in the field of architectural design, is the first female architect to win the Pritzker Prize, and has made remarkable contributions to the architectural landscape in China. Since the completion of the Guangzhou Opera House in 2010, Zaha Hadid Architects has undertaken 14 completed projects, including Beijing Leeza SOHO tower and Daxing International Airport, as well as 25 ongoing projects, injecting vitality and passion into our cities.

Guardian Art Center has always been dedicated to bringing high-quality art exhibitions from both domestic and international sources to its audience. The unveiling of the exhibition, 'The New World,' at the Guardian Art Center showcases Zaha Hadid Architects (ZHA)'s works from the past four decades around the world. It explores how ZHA continuously pursues innovation in architectural design through new digital design tools, robotics, 3D printing, artificial intelligence, and the concept of the metaverse. The exhibition also highlights ZHA's commitment to sustainability in the field of architecture and its renowned research achievements in environmental protection.

Hosting a grand architectural design exhibition in Beijing has been a long-standing desire of Zaha Hadid and the firm, which has finally come to fruition in collaboration with the Guardian Art Center. Taking this opportunity, Guardian Press and ZHA jointly present this title, of the same name as the exhibition, capturing the splendid moments of this architectural art exhibition and providing additional insights and continuations of the exhibition content.

Within the context of the new world, Zaha Hadid Architects responds to current challenges and envisions the future, expressing unwavering optimism towards what lies ahead. It is hoped that through The New World, viewers can freely navigate the urban fragments constructed by ZHA, refocusing on the relationship between humans and nature, contemplating the concept of existence, finding resonance, drawing strength, and redefining the future in a heartfelt way.

目 录

未来之境		1
采访帕特里克·舒马赫		9
采访大桥谕		23
展览		30
	"未来之境"展览介绍	32
①	扎哈·哈迪德建筑事务所序章	78
②	算法设计	150
③	机器人建造	166
④	数字社会	172
⑤	可持续发展	178
⑥	超级天际线	208
⑦	启程中国	212
⑧	对话北京	236
⑨	建筑未来	260
⑩	元宇宙隧道	358
⑪	扎哈·哈迪德建筑事务所经典作品年表	370

未来之境：扎哈·哈迪德建筑事务所设计展展区平面图

Contents

The New World — 5
Interview with Patrik Schumacher — 15
Interview with Satoshi Ohashi — 27
Exhibition — 31

    Introduction to 'The New World' Exhibition — 33
① Introduction to Zaha Hadid Architects — 79
② Zaha Hadid Architects CODE — 151
③ Robotic Assisted Design — 167
④ Zaha Hadid Architects Social — 173
⑤ Sustainability — 179
⑥ Super Skyline — 209
⑦ Zaha Hadid Architects in China — 213
⑧ Conversation with Beijing — 237
⑨ Building the Future — 261
⑩ Metaverse Tunnel — 359
⑪ Zaha Hadid Architects Timeline — 371

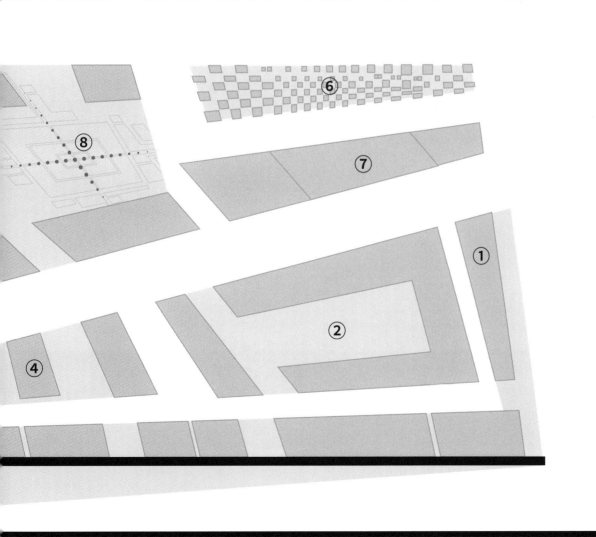

未来之境

帕特里克·舒马赫
扎哈·哈迪德建筑事务所 总裁
伦敦，2023年6月

"新世界"[1]也是汉内斯·迈尔于1926年撰写的一份激动人心的宣言的标题。迈尔是一位激进的现代主义建筑师，于1928年接替瓦尔特·格罗皮乌斯担任德绍包豪斯学校的校长。迈尔的宣言充满了对新世界的热情和期待：

> "福特和劳斯莱斯已经打开了城市的核心，消除了距离，抹去了城乡之间的界限。飞机在空中穿行……这些同时发生的事件极大地扩展了我们对'空间和时间'的概念，丰富了我们的生活。我们的生活节奏变得更快，因此似乎活得更久……在办公室和工厂工作的时间被精准划分成小时，铁路时刻表以分秒为单位计时，时间的概念在我们的生活中变得更加明确……无线电、电报和光电传真使我们突破了国界的阻隔，成为世界社区的一员……每个时代都需要自己的形式。我们的使命是用当下的新形式为我们的新世界赋予新形态……从理想和基本设计上看，我们的房屋正是一台活的机器。"

迈尔准确评估了他所经历的技术变革及其对社会生活的深远影响。他坚持这些变化必须在脱胎换骨的新建筑和城市主义中得到体现。这个从1920年到1975年的时间段，现在被称为福特主义时代。在这个时代背景下，工业化生产创造了前所未有的消费品，如汽车、洗衣机、收音机等，这也意味着工业化生产不仅使纺织品和陶瓷制品等传统产品的生产更加高效，而且最终融入并改变了城市生活。机器变成个人所有，不再隐藏在工厂中，这打破了19世纪城市和乡村生活之间无法跨越的高墙。诸如汽车这样的高端产品的大规模生产也意味着前所未有的工业集约。在这个时代，发达国家利用数量有限却普遍可得的商品确立了一种新的生活方式和一个统一的消费标准。标准化公寓就是其中之一，它的出现促成了空前的生活方式同质化。建筑的现代主义在勒·柯布西耶和汉内斯·迈尔等人积极、富有行动力的推动下，理解了这种新条件，并为建筑和城市主义得出了合乎逻辑的结论。现代建筑的先驱们根据他们大胆的愿景成功地开创了一个崭新的世界，彻底重塑了全球建成环境。现代主义因而成为第一个真正的国际风格。

然而，当扎哈·哈迪德在20世纪七八十年代进入建筑领域时，这种深深植根于大规模机械化生产技术时代的现代主义建筑和城市主义，在经历了50年的成功和霸权后，正面临着终结性的危机。在这个时期，大规模机械化生产的福特主义范式转变为基于微电子革命和计算机革命的后福特主义网络社会的新范式，并通过互联网进一步实现全球化、网络化和动态化（实际上，迈尔所预见的机械运输全球化潜力已被互联网和新社交媒体的出现指数级放大，并将很快由元宇宙进一步增强）。这一转变导致生产和消费方式出现了更大

---

1. "未来之境"的英文"The New World"可直译为"新世界"。

的差异化，进而形成了新的范围经济，摆脱了规模经济的单调重复。经济整体以迅猛的速度变得越来越具有复杂性和动态性，同时，这些新的特性也在城市面貌中逐渐显现。这些变化与现代主义所应对的那些变化一样，对社会生活和城市影响深远。科林·罗等建筑界人士以"拼贴城市"等概念做出回应；后现代主义和解构主义也以试探性、非决定性的回应出现，指向正确的方向。然而，我们应该为 21 世纪所准备、构想并当作我们学科的集体项目的是，全球建筑环境整体面貌的转变，该转变同我们在 20 世纪目睹的现代主义转变一样激进，却又截然不同。

　　扎哈·哈迪德始终将她的作品视为一个更大项目的一部分，她有鸿鹄之志，希望为构建一个更美好的世界做出贡献。1983 年，她精心设计并创作了一幅名为 The World（《世界》）的大型绘画作品，其中汇集了截至那时她所创作的所有建筑设计作品，包括她在 1982 年前后的里程碑式作品——香港山顶俱乐部竞赛方案。扎哈本能地把握住了后福特主义的新精神，并将其转化为建筑和城市设计中新的空间和几何复杂性。早在那时，The World 就展示了建筑师在应对城市化新动态和生产性社会生活演变时所需的新敏感性。

　　显而易见，我们的新世界不再是迈尔的机械世界，不再受到"我们在办公室和工厂的工作时间被精准划分成小时"的束缚。恰恰相反，这个世界属于自我决定和自我导向的富有创造性的工作，在不断变化的网络中，工作、休闲和持续学习融为一体。在这个新世界里，用户可以创造内容和产品，人工智能机器人取代了物理装配线工作，工业系统在吸收创新能力上几乎没有上限——为机器人或 3D 打印机重新编程，将新软件作为服务应用上传给全球用户已成为每天，甚至每分钟都在发生的事情。与常规生产相比，持续的产品、服务创新需要企业、行业和学科间更加密切的合作，同时融合科学、工程、设计、融资、营销等领域。这促进了沟通强度的增加和网络活力的发展，要求建筑在城市内进行创造性重塑。这个世界的关键词是差异化、多元性、多重关联、分层、相互可见和清晰度，与现代主义的功能分离、专业化和重复原则形成鲜明对比。在这个新范式中，一切都变得动态的相辅相成。

　　1983 年，扎哈·哈迪德就在对未来建筑和城市的设想中谈到了上文中的活力和流动性。然而，在此之后的 10 年里，她仍然是一位"纸上建筑师"，直到 1993 年维特拉消防站的竣工，世界第一次看到了我们打造新世界的决心。自那时起，我们最具开创性的大型项目皆在展示我们如何在日渐增加的复杂性中创造一种新的建筑秩序，并保持其可读性。北京的银河 SOHO 项目是我们的一个建筑典范，它的成功对我们此次展览中展示的众多复杂项目具有开创性的启发意义。2016 年，在蛇形画廊举办的扎哈·哈迪德作品展中，我们的 ZHA VR 团队开发了一个虚拟现实空间，重现了扎哈·哈迪德在 1983 年绘制的那幅画。在 1983 年，那仅仅是一种愿望和幻想，而在 40 年后的今天，在 2023 年，虽然分散在世界各地，没能构成自己连贯的世界，但已经是全球范围内一种实际的存在。或许我们在成都独角兽岛城市设计中证明了我们有设计整个城区并将其建设起来的激情和能力，然而，这样的机会是十分难得的。我们的想法从来不是单枪匹马地创造一个小世界，而是通过参与集体项目和思潮来影响更广阔的世界。

扎哈·哈迪德与解构主义的关联标志着一个重要运动的开始，它是对现代主义结束的早期回应，也是对未来的预期。从这种过渡性的解构主义风格中催生出了更宏大的参数主义运动。作为21世纪唯一的新兴原创风格，参数主义有能力成为定义未来数年的建筑方法、价值观和成功标准的划时代全球风格。我们的作品和此次展览证明了这种潜力。

在过去的20年里，参数主义通过广泛的研究和实践逐渐成熟。我将其当前阶段称为建构主义——不同于其他任何风格，它融合了精密的工程智能，并且进一步丰富了我们对日益复杂的社会生活进行空间化和表达的形态学选项。建构主义已在全球范围内崭露头角，尤其是在中国。中国引人注目的经济和城市发展已成为建立未来几十年的划时代建筑风格的主要舞台。我们对我们的作品在中国大规模地被接受和实施感到非常自豪和兴奋，因此对于在北京举办这次展览感到更加激动。

以大胆的姿态，我们将这次展览命名为"未来之境"，这反映了我们对建筑为人类进步做出贡献的乐观态度和持续追求。我们坚信，为了真正实现这一宏伟目标，全世界的建筑和城市必须以全新的理念和方法重新构想和设计。我设想中国将成为这一讨论和转变的开创性舞台，因此我非常荣幸地宣布，我的新书《建构主义：21世纪的建筑学》已于今年夏天以中英文两种语言出版发行。

新世界的建筑正在孕育之中。让我们一起将它实现。

The New World

Patrik Schumacher, London, June 2023
Principal, Zaha Hadid Architects

'The New World' (Die Neue Welt) is also the title of a thrilling manifesto written in 1926 by Hannes Meyer, the radical modernist architect who succeeded Walter Gropius as director of the Dessau Bauhaus in 1928. Meyer's manifesto brimmed with enthusiasm and anticipation of a new world:

> 'Ford and Rolls Royce have burst open the core of the town, obliterating distance and effacing the boundaries between town and country. Aircraft slip through the air...The simultaneity of events enormously extends our concept of "space and time," it enriches our life. We live faster and therefore longer...The precise division into hours of the time we spend working in office and factory and the split-minute timing of railway timetables make us live more consciously...Radio, marconigram, and phototelegraphy liberate us from our national seclusion and make us part of a world community...Each age demands its own form. It is our mission to give our new world a new shape with the means of today... Ideally and in its elementary design, our house is a living machine.'

Hannes Meyer rightly appraised the profundity of the technological changes he witnessed and their impact on societal life. He insisted that these changes must be reflected in a radically new architecture and urbanism. This era, spanning from 1920 to 1975, is now known as the era of Fordism. This is the era when industrialisation created a whole new unprecedented suite of consumer products such as automobiles, washing machines, radios etc, implying that industrial products went beyond merely making the production of traditional products like textiles and ceramics more efficient and finally entered and transformed city life. Machines became individually owned, no longer hidden within factories, thus impacting both the city and the gentile life that remained untouched during the 19th century. The mass production of sophisticated products like cars also meant industrial concentration at an unprecedented scale. The era established a new way of life and a universal standard of consumption in advanced nations, with a limited number of universally available products. Standardised apartments emerged as one of these products, contributing to a previously unseen homogenisation of lifestyles. Architectural modernism, fuelled by the vigorous and action-oriented energy and enthusiasm of figures like Le Corbusier and Hannes Meyer, understood the new conditions and drew logical conclusions for architecture and urbanism. The pioneers of modern architecture successfully and visibly established a new world, radically reshaping the global built environment according to their audacious vision. Modernism became the first truly international style.

However, this modernist architecture and urbanism, rooted in the technological era of mechanical mass production, faced a terminal crisis after fifty years of success and hegemony when Zaha Hadid emerged in the field of architecture during the 1970s and 1980s. It was during this period that the Fordist paradigm of mechanical mass production transformed into the new paradigm of the Post-Fordist network society, on the basis of the micro-electronic/computational revolution and then further globalised, networked, and dynamised by the advent of the internet. (In fact, the globalisation potential anticipated by Meyer due to mechanical transport has been exponentially amplified by the internet and new social media, soon to be further empowered by the metaverse.) This shift resulted in a much more differentiated approach to production and consumption, allowing new economies of scope rather than requiring monotonous repetition on a massive scale. The whole economy became increasingly complex and dynamic at a rapid pace. This new economic and societal complexity and dynamism began to manifest themselves in the physiognomy of the city. These changes and their impact on societal life and the city are no less profound than those which modernism had responded to. Architectural figures like Colin Rowe responded with concepts such as 'Collage City.' Postmodernism and deconstructivism also emerged as tentative, non-conclusive responses pointing in the right direction. However, what we should be preparing for, envisioning, and indeed making our discipline's collective project for the 21st century, is the transformation of the whole physiognomy of the global built environment—one that is equally radical but very different from the modernist transformation we witnessed in the 20th century.

Zaha Hadid had always seen her work as a part of a larger project, contributing to the construction of a better world. She dared to dream big. In 1983, she meticulously designed and executed a large painting entitled *The World*, in which she compiled the architectural designs she had created up to that point, including her seminal winning competition entry for the Hong Kong Peak in 1982/83. Zaha intuitively grasped this new spirit of Post-Fordism and translated it into a new level of spatial and geometric complexity, both in architecture and urbanism. *The World* painting serves as an early demonstration of the new sensibility required from architects to navigate the new dynamics of urbanisation and the evolving nature of productive social life.

It is evident now that our New World is no longer Meyer's mechanical world, bound by 'the precise division into hours of the time we spend working in office and factory.' Instead, it is a world of individual self-determination and self-directed creative work, in everchanging networks where work, leisure, and continuous learning merge. This new world enables user-generated content and products, where physical assembly line work is supplanted by AI-powered robots; it is a world where industrial systems, unlike rigid mechanical assembly lines, have nearly limitless capacity to absorb innovations—reprogramming the robots or 3D printing machines and uploading new software as service apps to a global audience becomes a daily or even minute-to-minute occurrence. Continuous product and service innovation, in contrast to routine production, necessitates a much more intensive collaboration within and across firms, industries, and disciplines, integrating science, engineering, design, financing, marketing, and more. This fuels the communication intensity and network dynamics, demanding architecture to creatively reshape cities and buildings within cities. Differentiation, mixity, multiple affiliations, layering, inter-visibility, and legibility are the watchwords here, standing in contrast to modernism's principles of functional separation, specialisation, and repetition. Everything becomes dynamically interdependent in this new paradigm.

Zaha Hadid expressed this dynamism and fluidity of her architectural and urban visions in 1983. However, for the next 10 years she remained a 'paper architect,' until the completion of the Vitra Fire Station in 1993 provided an initial glimpse of our determination to build the dynamism and complexity of the new world. Since then, our most seminal projects at scale have demonstrated our ability to create a new architectural order that could maintain legibility amidst increasing levels of complexity. The Galaxy SOHO project in Beijing is a built exemplar for us, a success that has been seminal in inspiring the multitude of complex projects showcased in our exhibition. In 2016, as part of the Zaha Hadid show at the Serpentine Gallery, our Zaha Hadid Virtual Reality Group developed a spatial VR version of Zaha Hadid's 1983 painterly vision. What was once a wishful aspirational fiction in 1983 has, to some extent, become a built reality forty years later in 2023, although dispersed across the globe rather than concentrated in one place to formulate its own coherent world. Perhaps our design for Unicorn Island Masterplan in Chengdu demonstrates our ability and passion to design a whole urban district and to see it through to construction. However, such opportunities are exceptions, and our idea has never been to forge a small world single-handedly but to impact the larger world by participating in a collective project and movement.

Zaha Hadid's association with deconstructivism marked the beginning of a significant movement, an early response to the end of modernism and an anticipation of things to come. What emerged from this transitional style of deconstructivism was the much bigger movement of parametricism. As the only new, original style of the 21st century, parametricism has the capacity to become the epochal global style that defines the methodologies, values, and success criteria of architecture in the years to come. Our oeuvre and this exhibition serve as a testament to this potential.

Over the past 20 years, parametricism has matured through extensive research and implementation. Its current phase, which I refer to as Tectonism, integrates sophisticated engineering intelligence unlike any other style, while further enriching the morphological repertoire of spatialising and articulating our increasingly complex societal life. Tectonism has gained prominence worldwide, with notable protagonists, particularly in China. The remarkable economic and urban development of China has become the primary arena for establishing the epochal architectural style of the coming decades. We take immense pride and joy in the reception and implementation of our work at scale in China, making it all the more thrilling to showcase our work here in Beijing with this exhibition.

With audacity, we have named this show 'The New World,' reflecting our unabated optimism and continued aspirations with respect to architecture's contribution to human progress. We firmly believe that the architecture and urbanism worldwide must be reimagined and redesigned with new concepts and methods to truly fulfil this ambitious role. I envision China as the seminal arena for this transformation and discourse, which is why I am personally delighted to announce that my new book, *Tectonism—Architecture for the Twenty-First Century*, will be published this summer in both English and Chinese languages.

The architecture of the new world is in its gestation. Let's make it happen.

采访帕特里克·舒马赫

扎哈·哈迪德建筑事务所 总裁

W 您好，很高兴再次见到您！第一次和您见面是在2010年SOHO中国举办的讲座上，那是一次向中国观众介绍银河SOHO和望京SOHO的精彩活动。我仍然保留着演讲的邀请函："帕特里克·舒马赫将为大家阐述：信息时代的城市空间是多元、开放和融合的。这个时代的建筑是多维的、自由的、参数化的，人们在这样的建筑中充分地拓展交流。受到建筑的影响，过去那种固定的关系——直线且平面化的思维习惯、生活方式都将彻底改变。"当时，您演讲的题目是"设计的未来"。时隔13年，扎哈·哈迪德建筑事务所（ZHA）仍然在塑造与定义城市未来空间的领域中遥遥领先。能否请您谈谈ZHA是如何进入中国市场，并成为在中国运营得最成功的国际建筑公司之一的吗？

P 很久以前，我们就进入中国市场了。最初，我们发现有一些竞赛可以参加，而且我们有中国员工，特别是来自香港的员工，他们对广州很熟悉。因此，我们在中国的第一座建筑就是广州大剧院，这就是一个竞赛项目。随后，我们通过SOHO中国的项目迈出了下一步，受邀为"SOHO城市"做了许多草图设计和提案，这个项目非常大，地点在北京的东三环。虽然最后的选址没有确定下来，但是我们规划了一个非常大的城市区域，有将近200万平方米，并且尝试设计了带有不同配置与构造的塔楼。那是一个非常棒的机会。突然之间，我们意识到，我们一直在思考的参数化城市主义，即如何生成带有全球秩序的城市，以有序的方式呈现良好的差异性，使不同分区能在保有各自特征的同时，兼具某种特定的动态结构和国际化程式——或许可以通过这个绝妙的机会实现。不过，遗憾的是，这个项目没有建成。此外，我们还应邀参加了位于北京市中心的"双子塔"竞赛。在竞赛中，我们将两座巨大的塔楼打开，让一个宽敞的中庭穿过两个塔楼，还有优美的地面环境——没有裙楼，让塔楼向外"绽放"，然而这个计划也未能实现。我们参加了这些竞赛，也赢得了竞赛，但最终这些项目未能落地。

之后，我们参加了北京银河SOHO的设计竞赛。那是第一次我们在北京不仅赢得了竞赛，深化了设计，还最终实现了项目。我认为那是一个很美的项目，并且至今仍然为它感到自豪。它展示了一个都市空间集群的理念，每个空间都有一个中庭，空间之间有很强的张力，体块之间有连桥相连，所以整个建筑的连接性非常强。项目有很多面向室外的空间，包括零售、餐饮与联合办公空间，除此之外，还有室内中庭空间和顶部办公室。不同于那些人们总是处在室内的大型购物中心，在这座建筑里，人们可以无障碍游走于室内和室外空间。现在，我们在那里有自己的办公室，也就是说，我们在自己设计的大楼里有自己的办公空间，一共有70名员工，这是非常美好的。此后，我们承接了第二个项目——北京的望京SOHO，以及第三个项目——毗邻上海虹桥机场的凌空SOHO。

近期，我们完成了拥有壮丽中庭的丽泽SOHO——我一直与SOHO中国的行政总裁张欣保持联系。我和我的团队在她的办公室里工作过一段时间，所以我对她和她的团队很熟悉。我跟她谈到过一个奇妙的想法——通过一个巨大的中庭来组织建筑物。那时候，我也是哈佛大学的教授，担任约翰·波特曼教授教席，约翰·波特曼在20世纪70年代便以这些梦幻般的中庭而闻名。我和张欣谈及了这些，之后，设计丽泽SOHO的机会就来了。我们提议："我们想做一个大中庭。"她说："好的，那我们开始吧。"

从那以后，我们就有了许多设计公共建筑的机会，例如，长沙梅溪湖国际文化艺术中心——一个坐落在美丽场地上的多场馆综合体，以及南京国际青年文化中心——一个拥有两家酒店的会议中心。最近，我们又中标了几个体育场馆设计的项目，其中一个是西安国际足球中心，已经接近完工。另外，位于成都的独角兽岛城市设计项目，我们负责设计那里所有的建筑，目前这些建筑都在建设中，我们已经进行到立面和细节设计的环节了，这也是最令人惊叹的项目之一。我们爱中国，在这里，我们可以做大规模的项目，可以设计一个城区，就比如正在成都进行的这个城市规划和城市设计项目。我们还设计了很多文化建筑，比如，在珠海我们有一个很棒的项目——珠海金湾航空城市民中心。这个项目在同一个大屋顶之下，有博物馆、音乐厅及各种不同的文化元素。最近，随着这些项目的成功，我们又获得了许多新的机会，包括一个歌剧院和博物馆的项目。在香港，我们也正在打造一座美丽的塔楼——恒基地产大厦。现在深圳有很多机会，我们也在那里设置了办公室，例如，OPPO国际总部和深圳湾超级总部基地C塔，还有深圳科技馆（新馆）。深圳一直是一个令人惊叹的城市，也是我最喜欢的城市之一，它充满活力。我们就是这样进入中国开展业务的。中国是我们最重要的市场之一，目前中国与中东是我们作为国际事务所最强大的市场。

W 您对ZHA如何进入中国以及如何拓展的历程回溯非常令人兴奋。众所周知，您于1988年加入ZHA，并在其成为全

汪芸→**W**
帕特里克·舒马赫→**P**

球500强建筑设计品牌的过程中发挥了重要作用；2016年以来，您一直担任公司的负责人。能否请您谈谈这些年来ZHA的发展和转型？

P 扎哈独立开创了这个建筑事务所。她曾在伦敦的建筑联盟学院（AA）学习，是一名优等生，毕业时获得了"荣誉文凭"。随后，她加入了她的老师雷姆·库哈斯和埃利亚·曾赫利斯的事务所——大都会建筑事务所。目前，大都会建筑事务所仍然是世界范围内最大的建筑事务所之一，他们因打造了中央电视台大楼而同样在中国名声大噪。扎哈和他们合作了一个项目之后意识到自己有着强烈的艺术个性，需要自由表达，而不是与合作伙伴协商设计。于是，扎哈成立了自己的事务所，参与了一些竞赛，但那时她还没有办公室，只能居家工作。同时，她也在建筑联盟学院教书，利用学校的场地、与学生共事。1986年，她雇用了一些员工并租下了她的第一间工作室。

我当时在伦敦上学，并且已经通过出版物了解到她。我曾经在德国学习，之后才搬到伦敦。1988年，我中断学业，加入了扎哈的工作室。当时，工作室只有一间半房间，5个人，我加入之后，我们成了6个人的团队。那时，我们参与了许多竞赛。除了两个日本的项目，我们手头没有工作，而那两个日本的项目最终也没有落成。在那个阶段，我们唯一实现的项目是维特拉消防站，它也是我的第一个项目。作为一个土生土长的德国人，我对这个德国项目有很深的感情。当时，我负责做细节深化，还非常缺乏经验，不过得到了经验丰富的建筑师的帮助，其中包括一位德国本土建筑师，这对我来说是一个突破。不过，在20世纪90年代的大部分时间里，我们仍在做非常小的项目，一些大的项目并没有落成。我们赢得了一些竞赛，但没能实现。尽管如此，这对我们来说依然很有意思，我们拓展了许多设计技能，并开始使用计算机。在我和扎哈共事的前10年中，我们没有完成任何大的项目，只有维特拉消防站、一个小型展馆及类似的项目，我们在20世纪90年代末才迎来真正进入主要建筑师阵营的转折点。

1999年，我们赢得了三项重大的竞赛，分别是位于意大利罗马的"MAXXI：国立二十一世纪美术馆"、德国沃尔夫斯堡的费诺科学中心和美国辛辛那提的罗森塔尔当代艺术中心。其中，罗马的项目是最重要的，它证明了我们不仅可以画出好看的图纸、举办展览和讲座，还能真正建造出美丽、不同寻常且复杂的建筑。那时，我的工作重点就是MAXXI。我全身心地投入，与团队一起开发这个项目，参与设计工作、每周例会，不断推进。这个项目是我们的突破口。不久之后，当我们正集中精力做这个项目的时候，世界各地的房地产出现了惊人的繁荣景象，特别是在迪拜和欧洲。这样的欣欣向荣一直持续到21世纪初金融危机的出现。

从2003年至2007年，我们同那个繁荣时期一起成长，事务所的规模在短短几年内从25人发展到450人。我们还很年轻，经验不足，手头的许多项目只停留在初期设计阶段。虽然很多项目都没有落成，但我们得到了大量经费，得以扩大团队并雇用更多员工来完成我们在迪拜的重要建筑设计项目，包括迪拜的歌剧院、股票市场和阿布扎比表演艺术中心，等等。仅在迪拜我们就拥有这么多项目。同时，我们几乎在西班牙的每座城市都有项目，在意大利、法国和德国也都有项目。所以，这是一次巨大的事业版图的扩张，特别是在迪拜和西班牙，但是金融危机的发生使许多项目最终停滞。好在那时候我们已经成长为一家规模非常大的公司了，并且有一些正在施工过程中的项目，例如，罗森塔尔当代艺术中心得以落成，接着，沃尔夫斯堡的费诺科学中心和MAXXI完工，而且扎哈获得了普利兹克奖。罗森塔尔当代艺术中心的完工和所有其他这些项目的推进使我们充满活力，帮助我们渡过了金融危机。扎哈获得普利兹克奖给公司带来了更多的机会。（2008年）金融危机之后，许多公司都缩减了规模，而我们还能继续在北非沿海地区开展许多项目。遗憾的是受到"阿拉伯之春"事件的影响，2012年以后其中一些项目无法继续展开。与此同时，欧洲主权债务危机也影响了我们在欧洲地区的一些项目。我们的状态略有下滑，但经历了这个时期之后，我们更加强大了。通过危机，我们变得更有策略，在人力资源和投资方面更加高效和谨慎，不会将过多的费用花在四处奔波以及其他不是那么重要的事务上。我们完善了规章制度，作为一家企业，变得更加成熟，这至关重要。

在2008年之后的几年中，我们经历了一个逐渐成熟和学习的阶段。近10年，我们已经成长为一个非常成熟的专业组织，拥有良性的公司结构、董事会和强大的财务部门。我们建立了正确的、经营一家大公司所需的所有架构，这是我们在过去10年里学会的。我们很有效率，可以快捷、高效地在预算内按时交付，并且在每个项目上都能赢利，发展非常稳健。我们实力雄厚，拥有良好的现金储备以及出色的履历和新的项目委托。在过去的几年里，我们的员工人数一直在增长，已经超过了500人。

扎哈去世之后，有那么一段时间，我们不确定会发生什么，不知道该如何渡过难关。但客户们非常忠诚，所有项目都在继续，也陆续有新的项目委托。公司的声誉已经建立起来，客户相信我和我的同事们会继承扎哈的遗志。其实当时我并不确定他们会不会继续相信我们，但这确实发生了。我们一直在出色地工作，以专业的方式交付，所以，到现在为止，我们已经很成熟了。扎哈刚去世时的慌乱几乎不复存在。我们爱扎哈，将她深深镌刻在心里，同时我们延续了她的遗志。我们能做到这一点真是太棒了。

W　确实如此。我之前没有意识到草创阶段你们所经历的充满动荡的历程，而这些危机使你们变得越来越强大。接下来，让我们把话题转向ZHA的核心。您经常提到技术与研究，众所周知，您是建筑、城市规划和设计领域最杰出的思想领袖之一。早在27年前，也就是1996年的时候，您就在建筑联盟学院成立了一个设计研究实验室。2013年和2018年，您还担任过哈佛大学设计研究生院的客座教授。那么，作为建筑实验的积极倡导者，能否请您谈谈ZHA的研究部门及其职能和定位？

P　教学对于开发新想法、测试新理念颇有裨益。当学生在一个项目上可以工作半年或一年以上时，我们就会有很多时间一起深化新想法。但是面对客户，在业界真实的竞争语境中，许多决策仅仅发生在数周之内。因此，提前构思并将人员引入公司进一步开拓、深化新想法至关重要。这样我们就可以加快进度，并将这些想法应用在设计竞赛中。这是一个关键的机制，但是在创新和研究方面，有些事很难通过不断流动的学生群体来实现，你必须一次又一次地对他们进行培训。

由此我意识到，对于更加宏大的研究项目——这样的项目其实并不少，我可以寻求与博士团队的合作，他们时间更充裕，也更加成熟。我们有一些为期三年的项目，也有研究经费和研究员，专门开展此类研究。其中的一项研究是使用代理模型对空间进行实时活动或使用情况的模拟，着重于交流、互动、相遇和对话，例如，它们受到空间和家具布置等不同安排的更多或更少的促进。还有一项研究，我们针对设计工具做了很多技术软件的开发，集成制造和机器人制造流程，并将其运用到前期设计的几何开发中，从而在后期建设阶段更加具有可行性。

为此，我们设立了算法设计研究组（ZHA CODE）和数字社会研究组（ZHA Social）。在很长一段时间内，我一直对企业组织和企业空间规划很感兴趣，因为我认为知识经济型企业和优秀的科技公司在各个方面都是优秀的创新者，它们的运营、沟通和合作模式也非常具有创新性，整个社会都可以从这些公司学到很多东西。因此，我开始关注这些公司，并和学生们一起开发像谷歌园区或Facebook园区这样的项目。之后，我们成立了一个公司内部研究组来处理大型企业总部和办公空间的项目，不仅使用特殊的分析工具来测量这类空间内部的连接性和采光情况，而且用生成工具创造多版本的平面布局。最近，我们又成立了另一个研究组——可持续研究组（Sustainability），在对事务所全部项目展开研究的

同时进行专业建模。此外，我们还成立了一个非常有趣的新的小组，他们专注于城市规划和项目规划的游戏化，凭借3D虚拟模型和模块化组件来支持各种参与式游戏。这些游戏还没有正式推出，仍在开发的过程中，我们致力于邀请终端用户和客户在一定程度上参与集体设计的过程，以此了解他们的偏好并建立社区。例如，在住宅、孵化器或企业项目中，不同的使用者可以找到彼此，看到彼此所占用的空间，甚至共享一些空间。这就是当下的研究。

我们在建筑联盟学院设计研究实验室与我的同事沙吉布·侯赛因一起做这件事，还建立了一个内部研究小组。我们从教学转向与博士合作、与专业研究团队合作。随着公司规模的拓展，我们可以负担得起更多的专业研究团队。将这种方式融入我们的工作非常重要，因为当参加竞赛或面对客户的时候，需要在很短的时间内做出回应，不可能从那一刻才开始创新。因此，你需要提前创建新的主题、拓展新的能力、构建新的工具，随时准备好应对新的挑战。这就是我们一直在构建的研究机制，我们会继续为之努力。未来，我们可能会成立一个城市化专家组，甚至会成立一个专门研究建筑幕墙的专家组。我们发展得越快，就越有能力支持这些具有研究能力的专家组。

W　请问您是什么时候开始在扎哈·哈迪德建筑事务所设立专项研究部门的？

P　我们规模最大、历史最悠久的研究组是ZHA CODE。它已经成立近15年了。ZHA CODE自主研发了设计生成工具，即将大量工程智慧集成到早期草图设计中的工具，由此，我们的工作方式更加能够跨领域融合。我们从计算机图形学、动画行业引入工具，然后将大量插件植入犀牛软件（Rhino）。事务所的同事们与合作研究人员组成了完整的生态系统，他们有机地融入研究人员的网络，联合开发我们的工具组。这个研究组大部分的时间致力于为项目提供专业的服务，同时不断展开研究。这个阵容强大的小组正逐渐演变为一个专业的项目交付组，负责提高几何形体的复杂度、容载量与合理性，通过算法系统的控制，节省成本和时间。这就是15年以来的发展历程。

W　ZHA运用机器人、算法和AI进行大规模建筑设计。请问，您如何定位人的需求？如何通过技术打造以人为本、量身定制的体验？

P　是的，当然，这并不存在矛盾。我们运用工具，正是为了实现更高层次的社会功能，这是我的措辞，你可以称之为以人为本，但我更愿称之为以社会为本的互动、活动、交

流场景。这不仅关乎个体及其感受，更关乎空间如何促进其中特定行为的发生，如在大学场景中，空间如何能够激发协作、合作、知识交流和集体学习等行为。具体来说，我们通过模拟互动行为，研究空间中不同类型的个体的聚集。我们积累了大量不同的样本。例如，在公司场景中，我们会关注不同部门、不同级别、不同职能的角色——无论他们是访客还是员工，我们会模拟他们如何以及何时会相遇，还有他们见面后的可能性，比如，他们在偶遇时是否有机会坐下来快速地交流信息。我们正在开发这些模拟项目，我们是这样用技术实现以人为本、以活动为中心的设计的。我们所考虑的不仅是建筑的物理构建，还有它们建立交互并促进交互的功能，以及它们对塑造实际社会生活的作用，这才是客户真正感兴趣的。只有当混凝土结构和建筑物能够刺激交流和互动时，客户才会对其产生兴趣。我们非常关注这一点，我一直在谈论关注社会功能，这是建筑师的职责，而技术功能则是工程师的工作，我们在技术层面的参与只专注于控制氛围敏感性和审美信号等细节。我的意思是，空间氛围会告诉人们将要发生什么，我所前往的是什么样的公司，我会经历什么样的体验，这是什么样的餐厅，我所进入的是什么样的零售空间，等等。他们需要感受到这一点，这是非常微妙的。如果你想给人一种品质感、一种与众不同的感觉，那么细节至关重要。是的，你需要避免重复，或过于常见的平庸，你需要在细节层面上有创造力。这就是我们要深入结构细节来控制这种美学结果的原因，但技术责任当然要由工程师来承担。

作为有思想的建筑师、教学者和该领域的领导者，我们应该担心和思考的不是建筑技术，而是设计意图、社会氛围以及交流框架。这就是我的研究重点，这些需要新技术的支持。假设我要测试一个空间的氛围、光照条件、反射强度、材质及空间的软硬度，那么我需要非常精确的建模、渲染技术，我甚至可能需要运用VR技术来让客户体验这个空间——让他们了解设计师的表达，也让设计师能够理解他们的建议。当然，工程师会模拟稳定性，而建筑师会模拟他们创造的体验。

W 这很吸引人，因为你以一种有机的方式组合了所有这些复杂的元素。那么，让我们谈谈可持续性和环保建筑。ZHA在建筑和环境可持续性方面处于领先地位。于2022年完工的碧哈总部采用了新一代技术，是一个令人兴奋的案例。中国设计学类核心期刊之一，清华大学美术学院的院刊，也曾发表了这个项目的相关研究。能否请您进一步展开谈谈ZHA在这方面的探索？

P 我们希望对自己的工作成果负责。最近，我们成立了一个可持续研究组，因为我们讨论可持续发展和环境问题已经有很多年了。近年来，这个问题的紧迫性有所加强，许多公司致力于零碳或碳中和，并引起了颇多关注。因此，我们也提高了参与度。我们成立这个研究组，利用内部资源支持这些研究，同时，我们也在对过去的建造活动进行可持续评估。对于迄今为止已经建成的项目，我们委托工程师测量这些建筑的碳足迹和可持续性绩效，以便更准确地了解我们需要注意的事项。我还认为，重要的不是只关注碳足迹的最终测量值，因为人们可以通过使用当地劳动力、材料，使用特定机械设备或特定再生混凝土等许多方法来实现这类目标。更重要的是，建筑师可以具体做些什么，不止于技术系统层面，他们可以通过建筑的形态学以及在太阳朝向和风向方面的因地制宜来发挥作用。我们已经做了很多这样的项目，事实上，当你看到这些项目的时候，就可以直观地感受到它们的与众不同。我们还可以观察建筑东、西、南、北各方向的处理方式，来了解自己所处的环境。

这样的建筑运用了自然通风和自遮阳等多种被动式系统，例如，brise soleil部件（一种遮阳系统，通过在幕墙上开孔或设置叶片来控制进入建筑的自然光线和太阳能热量），当阳光从南向各方向移动的时候，这些部件产生的效果也会随之发生变化，使被动式系统更加有机和完整。同时，它们也改变了建筑的外观，你可以看到或者感受到这是一座可持续性很强的建筑，能高度适应气候条件。在实际项目中，类似的案例包括伊拉克中央银行，利用自身结构形成的阴影遮阳是这个建筑的基础；阿卜杜拉国王石油研究中心，这个建筑有许多小窗户、遮阳庭院和顶棚，它们被有意识地设计成朝北的方向，以配合风的方向。这些对我们来说都是具有开创性的项目，它们展示了有环保意识的设计也会让建筑产生新的外观。其实，比被动式系统更妙的是当地的建筑条件，因为我们在谈论的是被动式系统，人类对这种系统早在空调出现之前的那个时代就有所运用。直觉上，这些建筑物属于它们所在的区域，因为它们是根据当地气候量身打造的，而且很多时候，你根本不会注意到一个具有高度可持续性的建筑与其他建筑有什么不同，这是因为此类技术都被隐藏在某些机械设备和决策当中，尽管它们并非真正意义上建筑学的决策。这是我个人的偏好。

我还想说，可持续性原本是一种限制，只是在我所称的环境构造论中变成了一种驱动力。不过，我们需要持续关注建筑环境的品质，这是我们自始至终的纲领。很多时候，可持续发展只是纸上谈兵，因为单纯谈论它很容易，你可以量化它，你可以说，我有这些环保认证，所以我是一个好人、一个有责任心的人。但是，人们对于体验和互动的品质的关注和讨论是不够的，而这才是首要的。你希望塑造什么样的生命进程，一座建筑应该有多吸引人、多开放，而某些空间需要有多隐蔽——你首先需要考虑的是这些问题，然后再综合环境的限制考虑全局，不应该出现"尾巴摇狗"（即本末倒置）的状况。一些被视为限制的事物也有可能成为某种动机，但它永远不该是建筑或城市形成的主要原因。这就是我对可持续发展所持有的态度。我有一本以建构主义为主题的新书已经出版。在那本书中，我探讨了不同工程领域的进步，探讨了结构工程以及环境工程如何以新的方式塑造建筑环境与建筑物，尤其是当你感知到，哦，是啊，这是21世纪，这是可持续的，因为采用了这些被动式系统，就不需要消耗那么多的能量了。书中，我专门从可持续性这个角度探讨我的建筑，这应该是建筑学科的研究领域，而非基础设施、工程学或材料科学的。

W 这是您最新的出版物？

P 是的，这是一本新书。它已于2023年7月面市。事实上，它有中英文两个版本。现在就已经可以订购了，书名是《建构主义：21世纪的建筑学》。在书中，我专门探讨了建构主义与环境相关的内容；当然，也探讨了新的机器人建造及新的结构优化技术带来的影响。我认为这些都为建筑提供了美好的支持，有意思的是，这些使建筑看起来更具自然形态、更像有机体，这是因为自然界确实也存在对结构优化的适应性反应，自然界的万物都在进化中得到了结构上的优化，就像我们对自身所处的自然环境的适应。比如，一些近乎有机体的传统住宅并非刻意设计的结果，而是与所在地区的气候条件和物质资源磨合的必然结果，它们已经进化了数百甚至数千年。当代需求可能更为苛刻，因为我们必须建造体量更大的建筑，所以建筑物并不是对环境的直接反应。不过，我发现，有意思的是，我们的建筑看起来更像是源于自然，并且可能还以某种方式与传统建筑的复杂性、不规则性及规则性变化相联系。

W 您的解释让我对可持续发展有了更深的理解，非常有启发性。此外，很高兴了解到您的新书。在即将于北京嘉德艺术中心举办的展览"未来之境"上，将会展出ZHA的全球精选项目，包括120多个模型和许多视频投影，展览的最后还会以一个元宇宙隧道作为结束。您曾经谈到："我愿为之奉献的元宇宙，能够支持生产性社会生活，甚至成为社会生产和社会再生产不可或缺的一部分。这种严肃的元宇宙，并非现实替代或第二人生，也不是对社会现实的逃避，而是会促进社会的发展，元宇宙能够实现充实且富有成效的生活。"能否请您谈谈元宇宙带来的新的可能性与机遇？

P 谢谢。你找到了完美的引文作为这个问题的前提，清晰明了。目前，元宇宙的项目多为电子游戏——通过玩游戏赚钱，迎合青少年逃避现实、不承担责任的想法。我不想与此有太多关系。我们目前所做的是，从这样大型的多人游戏环境，即可以交互的虚拟环境中，进行技术转移。实际上，我们用这项技术所做的是将其与我们的日常生活联系起来，尤其是生产、文化、教育生活。所以，我认为你选择了正确的引述。目前，我们正在研究元宇宙，特别是面向我们最了解的群体展开对话，包括我们的同事、其他建筑师、室内设计师、服装设计师、汽车设计师、产品设计师、平面设计师、元宇宙设计师等，并邀请他们进入一个新的社区。在这里，你可以设置自己的商店，展示你在做什么，也可以进入其中，开会、讨论问题。目前我们正在我们的空间里举办虚拟的威尼斯建筑双年展"都城乌托邦"，而这仅仅是个开始。目前，我们已有一座主体建筑和一些其他的建筑，我们想把它建造成一个完整的虚拟设计区，甚至可能是一座虚拟城市。对我们而言，这很显然与我们的现实世界相互竞争，相互替代，紧密联系。（在现实世界中）会有进入这些虚拟世界的窗口。我们也会对虚拟世界进行类似的设计。有时候，会有这样的客户——他们既需要物理空间设计，也需要元宇宙空间设计，抑或是将两者相互结合的空间设计。这将是一种持续的社会现实。我认为这是一个非常重要的信息。作为建筑师，我们的核心能力在于创造用户体验，而不是创新技术。所有的物理工程界限将消失，设计概念会在元宇宙和物理空间之间实现一对一的转译。当然，在元宇宙中，会有一个全新的工程堆栈、一组新的限制与要求、多边形计数与刷新率，以及许多其他的限制，因此设计是相通的。如果你是一名建筑师，那你在设计元宇宙空间时仍然是一名建筑师，你并不会因此而偏离你的行业，而是在继续，因为这两个领域在实用性方面有着相同的要求和标准。在元宇宙空间里定位容易吗？你能辨别方向吗？我们能否在空间里找到彼此？我们可以愉快地参加讲座，或者亲密交谈，或者参观展览吗？我们可以与面前的3D模型讨论设计吗？关于成功的空间的所有标准都是相同的，无论物理空间还是虚拟空间，因为我们希望拥有相似的东西，实现相似的东西。我们希望见面、参观展览、在社交场合结识新朋友；我们对生活的要求是相似的，现在我们只是有了两种选择来满足这些要求，包括Zoom在内的社交媒体就说明了这种状况。不过，我相信元宇宙是因为我认为这种体验比其他远程方案更具吸引力。

在虚拟城市中漫游并社交，与写电子邮件、微信聊天以及参加虚拟会议是不一样的，甚至与Zoom会议也存在差别。我刚刚给中国听众做了一个线上讲座，我不知道究竟有多少人参加，因为观众只是视频界面中非常小的头像和名字，我不能和他们互动，不能将他们拉到一边闲聊，也不能在活动前后进行一些非正式谈话。所有的这些都不会在像Zoom这样的工具中发生。但是如果你在元宇宙中进行类似的讲座，这样的事情就能发生。你可以早一些到，看看有谁在那里；你可以做你的演讲，或者聊天、四处看看，更具互动性；你还可以看到大楼里发生了什么。或许你是因为要参加一个讲座而来到这里，但是你会留意到，"哦，那里发生了什么？有人正在解释这一款模型，让我迅速浏览一下并加入其中"，你能看到更多的信息，可以导航、浏览。而在互联网上，你并不能那样做。你在Zoom上也看不到其他同时发生的事件，只是一个接一个的事件。而在元宇宙的虚拟世界中，就像在现实世界中一样。你前往一所大学，会特别关注一些东西，但你从眼角能瞥到许多其他正在发生的事情，你可以切换关注点，等等。我认为这些优势非常吸引人，可以将其与真实的会议更好地结合起来——可以是几个人在一个有大屏幕的空间里开会，其余的人通过视频加入，而不总是一个人孤零零地坐在家里的桌旁对着显示器，可以同时将物理现实社交和虚拟社交整合在一起。这些优势对我来说意味着每个个体、组织、网站、建筑物、公司、慈善机构、大学、媒体机构，都将在虚拟城市中拥有虚拟空间。我了解这一点是因为我认为这种体验很吸引人。问题是，它什么时候会发生？当参与者寥寥的时候，就不值得去元宇宙了。那么谁先来？这就需要第一个吃螃蟹的人。正如早期的互联网，我记得当互联网刚刚出现时，它并不有趣，也没有多少网站。在谷歌出现之前，当你尝试搜索信息的时候，你会发现找到的并不是你想要的。所以，你会说，好吧，这不是我需要的。但是，随着不断的积累，达到阈值，网络效应就开始了。突然间，越来越多的人参与进来，然后一转眼，每个人都必须掌握它。我知道这一刻即将到来，我们正在为此做准备。我们所有人都即将迎来大量设计元宇宙空间的工作。这就是为什么我们需要配备人工智能和生成工具来完成这些工作，我认为这是关键。很高兴这即将到来，不过，尚且不清楚具体会是什么时候，因为目前可探访的元宇宙空间并不多。或许三年之后，每个人都会到访元宇宙。这是一个非常值得认同、接受的积极愿景。

W 太棒了。接下来，是我的最后一个问题。让我们回到未来这个话题。在您看来，建筑师需要为未来的知识型、创意型社会而设计，为能激发想象力的互动空间而设计。您关注科技公司如何构建它们的空间，如何选址，以及他们期望什么样的氛围。这也是您与建筑联盟学院和哈佛大学设计研究生院的学生一直在做的研究。按照您的解释，这样做，"文明会随之进化，社会的每一名成员都会过上更好的生活"。从建筑师的角度来看，您认为未来人们更好的生活是怎样的？或者说，您的愿景是什么？

P 我相信技术是惊人的，包括机器人、软件、服务、3D打印技术，这意味着生产力。从事农业的人会变少，因为会有机器人拖拉机负责播种，会有配备LED灯照明、能够进行25次收获的集约化地下农业。所有这些都是在现在的城市、技术中心、农业科学中心等地方开发出来的。我认为大多数人都会参与技术的开发、创新和再创新。其好处在于，如果我们做得正确，就会有许多收益，因此我们可以减少工作量，或者选择能激发自己兴趣的工作。同时，我们会持续学习，我们的工作会更有自主性。没有人会控制我们去做具有创造性的工作。我们能够保持自我驱动。疲倦的时候，我们可以选择去睡觉；当有其他事情要做的时候，我们可以自由支配自己的时间。同时，我们可以保持联络。这就是更好的生活。更好的生活是更富有成效、更没有限制、更灵活自由的生活。我认为，作为人类我们并不追求偷懒，也不享受无所事事和无聊。这就是为什么我认为这样的生活会更好，因为当我们富有成效的时候，我们会感到快乐。而这里所谓的富有成效指的是无须顶着超强的压力，也没有其他人指手画脚的成效。这就是我的愿景，为此，我们需要非常开放的空间，我们需要运用元宇宙。即便在像我们这样的公司里，我们也已经有了这种感受。你可以从不同经济体之间的差异感受到——指令和控制越少，就会有更高的参与度并产生更多自我激励性的工作。我认为，工作对愉悦感的产生很重要，只要这份工作不是苦差事，而是振奋人心的。这就是我的愿景，也是我对这次采访的结语。

W 非常感谢您向我们展示了扎哈·哈迪德建筑事务所的发展轨迹。我们谈论了可持续性、研究与中国。期待您的新书、新想法和新建筑。

P 非常感谢。很高兴见到你，再见。

Interview with Patrik Schumacher

Principal, Zaha Hadid Architects

Wangyun → **W**
Patrik Schumacher → **P**

W Hello Patrik, it is wonderful to meet you again! The first time I saw you was at your lecture organised by SOHO China in 2010, an exciting event to introduce Galaxy SOHO and Wangjing SOHO to the Chinese audience. I still have the invitation to your lecture with the following message, 'As Patrik Schumacher will elaborate, to accommodate the multi-faceted, open-minded and blended city space of the Information Age, the architecture and design of this era will be multi-dimensional, free, and parametric, permitting the complete extension of human communication. Unlike the rigid relationships, straight lines, and flattened ways of thinking, encouraged by the architecture of the past, this new architecture induces a complete change in our way of life.' The title of your lecture was 'The Future of Design.' It is fascinating that after 13 years, ZHA is still in the leading position to shape and define the future of our city. Could you please talk about how ZHA entered the Chinese market and became one of the most successful international architecture firms operating in China in a globl context?

P It was quite a while ago that we joined the market in China. We discovered that there were some competitions we could enter—we had some Chinese members of staff, particularly from Hong Kong, who were also familiar with Guangzhou. Therefore, our first building was the Guangzhou Opera House, which was a competition. The next step was actually through SOHO China, that we were invited to do a number of sketches and proposals for a very big project, so-called 'SOHO City,' on Beijing's third ring road. The site, in the end, was not secured, but we sketched a very large urban field with…I think it was nearly 2 million square metres, and we tried to design different fields of towers with different configurations. That was an amazing opportunity. Suddenly, we realised we had been thinking about parametric urbanism—how we could generate urban districts, which have a global order and have a nice differentiation but in an ordered way, so that you can have an identity for a district and a certain dynamic structure and global form. That was a great opportunity. Sadly, it did not happen. We couldn't do it. We were invited for another competition in the centre of Beijing for 'Twin Tower,' where we have these large towers opening up and allowing a big atrium to come through the two of them and a beautiful ground condition—no podium, but the towers flaring out. That didn't happen either. We did these competitions, and we won them, but the projects didn't come through.

We then joined the competition for what we now call Galaxy SOHO in Beijing. That was the first project in Beijing which we not only won the competition but also developed the project, which finally got realised. I think it is a beautiful project, and I am still very proud of it. It shows the idea of an urban cluster with a lot of intensity in between the spaces. Each of the volumes has an atrium with bridges connecting the volumes, so it is super strong on connectivity. The retail spaces also have a lot of exterior, outdoor retail, food and beverage, and co-working spaces, as well as indoor atriums, with offices on top. It was not these big malls where you are always inside, instead you can step in and out. We have our own Beijing office there now with 70 people in our own building, which is very beautiful. We then had the Wangjing SOHO in Beijing and a third project close to the Hongqiao Airport in Shanghai.

More recently, we completed the Leeza SOHO Tower with the beautiful atrium—I have been communicating with Zhang Xin, the CEO of SOHO China, and I was working inside her office with my team for a while, so I am quite familiar with her and her team. I was talking about this fantastic idea of a big mega atrium hollowing out buildings. At that time, I was also a professor at Harvard. I was a John Portman chair because John Portman is famous for these fantastic atriums in the 70s. Therefore, I was talking to her, and then the opportunity with Leeza Tower came up. We said, 'I want to do a big atrium.' And she said, 'Yes, let's do it.'

Since then, we have had so many great opportunities with other public buildings, like the Changsha Meixihu International Culture and Arts Centre, a multi-venue complex on a beautiful site, and Nanjing International Youth Cultural Centre, a conference centre with two hotels. We have done so many projects. Recently, we have also been winning quite a few stadiums, and one of them is nearly finished in Xi'an [Xi'an International Football Centre], as well as nearly the whole city district in Chengdu, Unicorn Island Masterplan, where we are designing all the buildings. They are all under construction, and we are doing the façades and detailing. This is one of the most amazing projects. We love China because we can do things at scale. We can do a city district, like we are doing a lot of masterplanning and city planning. Sometimes we can get big chunks of this, like in Chengdu. And we also do a lot of cultural buildings. In Zhuhai [Zhuhai Jinwan Civic Art Centre], we have a nice project with a museum, concert hall and different cultural components under the big roof, etc. New opportunities came up recently after we have won a number of projects—another opera house and another museum. We are also doing a beautiful tower in Hong Kong, The Henderson, and we are opening a Shenzhen office because there are so many opportunities in Shenzhen now, for instance, OPPO headquarters, Tower C at Shenzhen Bay Super Headquarters Base, as well as the Shenzhen Science and Technology Museum. Shenzhen has been amazing, and it's one of my favourite cities with a lot of dynamism.

This is how we got into China, how we expanded and are now expanding further. It is one of our most important markets, together with the Middle East at the moment, these are the strongest markets for us as a global brand.

W It's fascinating to hear you talking about how you got getting into China and how you expanded, so exciting. As we all know, you joined in ZHA in 1988 and have played a significant role in developing ZHA to become a 500 strong global architecture and design brand; since 2016, you have been leading the firm. Could you please talk about the development and transformation of ZHA during these years?

P Zaha started by herself. She was studying at the AA [Architectural Association School of Architecture] in London. She was the top student and won the diploma prize when she graduated. Then, she joined her teachers, Rem Koolhaas and Elia Zenghelis, at their company OMA [Office for Metropolitan Architecture], and they are still one of the fantastic big companies, which has also made a mark in China with the CCTV building. She was partnering with them for one project, but then she realised that she had her very own strong artistic identity. She needed the freedom to express herself and not negotiate her design with partners. As a result, she founded her own studio, and she did some competitions without having an office at that time. She worked from home, but she was teaching at the AA and working with some of the AA students and spaces. In 1986, she founded the office, hired some people, and rented her first room.

I was a student at the time in London, and I already knew about her through her publications. From studying in Germany, I moved to London and became a student here. I was still a student in 1988, but I interrupted my studies to join her studio, which was still just one and a half rooms with five people—perhaps six after I joined. We started doing a lot of competitions, but we didn't have work, except for the two Japanese projects, which didn't happen in the end. Then the only project which would happen at that stage was the Vitra Fire Station. It was my first project, for which I did a lot of the design. Being of German origin and fluent in the German language, I had a distinct German connection with this project. Although I was very inexperienced still, I was working on the detailing and had help from more experienced architects, one of them was a German local architect—that became my breakthrough for that building. But then for most of the 1990s, we were still doing very small things. We had some bigger projects which didn't happen, and competitions we won weren't realised, but still it was interesting and fascinating because we developed a lot of design skills, and we also started using the computer. I was working with Zaha without a major project for nearly over 10 years. We did Vitra and another small exhibition pavilion and such projects, but the breakthrough into the major league of architects really came towards the end of the 1990s.

In 1999, we won three major competitions: the MAXXI: Museum of XXI Century Arts in Rome, Italy, the Phaeno Science Centre in Wolfsburg, Germany, and the Lois and Richard Rosenthal Centre for Contemporary Art in Cincinnati, USA. The three significant projects, with Rome being the most significant, not only established us as designers capable of doing beautiful drawings, exhibitions and lectures, but also demonstrated our ability to develop remarkably beautiful and complex architecture. My focus at that point was the MAXXI, so I was personally focusing my energies on developing this project with the team, designing and going to meetings every week and building it up and getting it underway. That was the breakthrough, then soon after that, as we were working on this, we had this amazing real estate boom around the world, particularly in Dubai, but also in Europe and Spain and everywhere, which continued from the early 2000s all the way up to the financial crash.

From 2003 to 2007, we were growing together during that boom period, and we grew from 25 people to 450 people in just a few years. We were still very young and inexperienced, and a lot of the work were just early-stage designs. Many of them didn't happen, but we got a lot of fees so that we could expand and hire people to do all these major, major building designs in Dubai, such as Dubai Opera House and Cultural Centre, Dubai Financial Market, Abu Dhabi Performing Arts Centre, etc. There were so many in Dubai alone, and we also had projects in almost every city in Spain; additionally, we had projects in Italy, France, and Germany. This was with a huge expansion of Dubai and the Spanish projects in particular, but the financial crash stopped a lot of that.

Nevertheless we, at that time, had grown to a very large company, and in the meantime, we had some buildings which were progressing on site: like Cincinnati got finished, then Wolfsburg was finished, and MAXXI as well, and Zaha got the Pritzker Prize. With Cincinnati being finished and all these other projects underway, we were so dynamic at the time, which helped us through the financial crash. Zaha winning the Pritzker Prize brought us more opportunities to the company, and since the crash, what's interesting was a lot of companies had lost a lot of people, but we were able to keep going with a lot of projects in North Africa, all along the coast. The Arab Spring came in and stopped some of this, unfortunately, in 2012, but also the European debt crisis stopped some of the European projects. We were dipping a little bit, but then we came out of this much stronger. Through the financial crisis, we learned how to be more business-savvy, to be more frugal with our resources, to save projects, to invest more carefully in the resources, and not to spend so much of our fees on flying around and doing all sorts of things. We were more disciplined, and we matured as a business, and that was very important.

In the years since 2008, there was a maturing and learning curve, and then the last, I would say, 10 years, we have been a very mature professional organisation with a very good corporate structure, with the board of directors, a very strong finance department, and we built all the structures you need to run a large company properly, which we learned only in the last 10 years. Now we are very efficient, we can produce very fast, efficiently, on budget, on time, and make profit on every project and be very sound. We are well-established, we have good cash reserves, a fantastic portfolio and order book of new work. We have been growing in the last few years to over 500 people.

Since Zaha's passing, for a moment, we were unsure what would happen, how we can survive this, but the clients were very loyal, all the projects were continuing, and we also got new projects. The reputation of the firm was established, and they were trusting me and my colleagues to continue the legacy of Zaha, which is something I was not sure would happen, but it did. We kept delivering professionally. By now, we're well-established, and this initial worry after Zaha's passing is nearly forgotten. We love Zaha, we like to keep her memory, but we can continue her legacy. That was fantastic that we could do that.

W That's great. I didn't realise this was a journey full of turbulence at the beginning, and then you're getting stronger and stronger through all these crises. Let's move to the core of ZHA. I heard you always mentioned technology and research, and as we all know, you're one of the most prominently thoughtful leaders within the field of architecture, urbanism, and design. 27 years ago in 1996, you founded a design research lab at the AA. You also acted as a guest professor at Harvard GSD in 2013 and 2018. So, as an active proponent for the experimentation in architecture, could you please talk about the research department at ZHA, its function and position?

P The teaching helps a lot to develop new ideas, test out new ideas first, and we have more time to develop ideas with students when they work on a year or half a year at least on a project, but in the business or in the competitive world of real competitions and clients, it's only a few weeks. That is why it's very important: to develop the ideas there first and bring the people into the company and then develop what we've discussed and developed and realise it. We can then accelerate and bring these ideas to competitions, etc. That's an important mechanism, but there are some things which, on the side of innovation and research, are difficult to achieve with the student population, which is continuously changing, you have to train again and again.

I then realised for some more ambitious research projects, there's a number of them, I went to use PhD groups because they have a bit more time, they're more mature, and we have a three-year program and some funding with research fellows, to develop some of the research. One of them is the live process or occupancy simulations with agents, focusing on communication, interaction, encounters, and conversations, for instance, as they're facilitated more or less by different arrangements of the spaces and furniture situations, etc. There is one research, for which we do a lot of technical software development in terms of design tools but also integrating fabrication and robotic fabrication processes, and integrating that into early design geometry development, so that the geometry is downstream constructible.

For that, we have a separate group [ZHA CODE]. We also have a specialist group [ZHA Social]. I was, for a long time, interested in corporate organisation and in corporate space planning because I believe in businesses, in particular, knowledge economy business and great tech companies are great innovators on all fronts, and also the way they operate and communicate and work together is quite innovative, so the whole society can learn a lot from the example of these tech companies. I was focusing on them with students to develop things like Google campus or Facebook campus, and then we developed an in-house research group on coping with large corporate headquarters, and this idea of office spaces and using special analytic tools to measure connectivity within these spaces, light penetration, but also generative tools to generate layout versions, etc. That is another research group, and recently we also have developed yet another group—a sustainability group, with whom we do both research and specialist modelling for all our projects, and we finally have a very interesting new group, which is about the gamification of urban planning and project planning, supported by 3D virtual models and modular components that allow all kinds of participatory gameplay. These are not rolled out yet, but we're developing that also in the attempt to invite developers to allow their end users and clients to get drawn into a collective design process to some extent, to understand preferences and to build a community. For instance, in the residential project or in an incubator or corporate project, the different tenants can find each other, see each other taking up space, locating next to each other, maybe sharing some space. That's the current research.

We do that at the AADRL [the Design Research Laboratory] with my colleague Shajib Hussain who is doing this, and then we built up an internal research group. We are working from teaching to PhD to professional research teams. As we grew larger, we can afford more of these professional research teams, but it's very important to feed into our work this way because when you are faced with a competition or client, and you need to respond within a very short period of time, it's not possible to start innovating at that moment. You need to build up the themes, the topic, the capacities, the tools to then be ready to respond to new challenges. That's what we've been building up, and we continue to build up. We're discussing an urbanism specialist group, for instance, a specialist group for façades as well. The more we grow, the more we can afford these specialist groups, which also have research capacity.

W   That's great. But when did you start to have this specific research department at ZHA?

P   The biggest and oldest is called ZHA CODE, and I think this is probably nearly 15 years old already. We started to become more sophisticated after they have incorporated generative tools and integrated a lot of engineering intelligence into early sketching tools, which we developed ourselves. It's learning—bringing in tools from computer graphics from the animation industry with a lot of plugins into Rhino, where there's a whole ecosystem of colleagues and co-researchers, who are not only developing our own tool sets but also plugging into this network of researchers. That's the oldest strand of research, which is also quite large. At this point, they are, most of the time dedicated already to being specialist service in many projects. We're still researching, but this group is quite large and now becoming a specialist delivery group for increasing the geometric intricacy, capacity, and rationality, which is also very good for cost-saving, time-saving—using these complex geometries to get them fully under control with these algorithmic systems. That's 15 years in the making and has recently been professionalised.

W   And as we all know, ZHA builds things on a vast scale with robots, algorithms, and AI. How do you position human needs and create a human-centred, tailored experience with the technology?

P   Yes, of course, there's no contradiction. We use tools to precisely home in on a higher level of social functionality. That would be my phrase. You can call it human-centred. I call it more like social-centred interaction for the activities and the communication scenarios. It's not the individual humanbeing and what they might feel, but it's how the space is conducive to generate collaboration, cooperation, knowledge exchange, collective learning in a university setting, etc. So precisely, because we have these simulating interactions, we think about the different types of individuals who would come together in a space. We have a population of agents which are quite different. For instance, in a corporate setting, we would look at different departments, we look at different hierarchy and status levels, different functional roles, whether they're visitors or members of staff, and then we play out the scenarios of how and when they might meet and what opportunities they have if they, say, run into each other and encounter, to sit down and have a quick exchange of information. We're developing these simulations, and that's my way of bringing technology to a more human and activity-centred approach and not just thinking about the physical build, but we need to think about the building always as something which is framing the interactions and facilitating the interaction, an actual life process of society and what the client is really interested in. The client isn't interested in looking at a concrete structure or a building unless that stimulates communication and interaction and facilitates the process. We're focusing on this very much, and I've been talking a lot about focusing on social functionality—that's the architect's job, technical functionality is the engineer's job. We get involved in technical detailing to the extent that we would like to control the atmospheric sensitivity, the aesthetic signalling. I mean an aesthetic ambience will tell people what to expect, what kind of company I'm going to, what kind of experience I'm going to here, what kind of restaurant is it, what kind of retail space am I entering. They need to sense that, and it's very subtle. The details matter a lot if you want to give a sense of quality, but also you want a sense of difference. Yes, and you want to overcome the banality of repeating what's happening everywhere. You need to be inventive on the level of detail. That's the reason why we're also going into the detail of construction to control this aesthetic outcome, but the technical responsibility is of course with engineers.

What we, as thinking architects, teaching architects and leaders of the field, should worry about and think about is not the construction technology so much, but the design intent, the social atmospheres, and communication framings we're generating. That's what my research is focusing on. But I need new technologies. For instance, if I want to test the atmosphere and the lighting conditions, reflection, materiality, sense of softness or hardness of a space, I need to have very good rendering technology. I need to get very good modeling and precise technology and rendering and then I need maybe even to have a VR technology to explain to the client what he would experience when he's entering the place. So, the technology is precisely the means to allow the client and end user to get a feeling of what they should expect, and it also allows the designer to understand what they're proposing because they can simulate the final spatial experience. Of course, engineers will simulate stability, architects will simulate the experience they're trying to generate.

W   It's fascinating because you combine all these complicated elements in such an organic way. And so, let's talk about the sustainability and green architecture. As we all know, ZHA is in the leading position in making the building and environment more sustainable and green. The BEEAH Headquarters completed in 2022 with the next generation technologies is an exciting sample. The school journal of the Academy of Arts and Design at Tsinghua University, one of the leading academic design journals in China, is publishing an article on this project. Could you please talk more about ZHA's exploration in this area?

P   We want to take responsibility for our own production. We recently established a specialist task force on sustainability because we've been talking about sustainability and environmental issues for many years—many, many years. In recent years, there was a ramp up of urgency, a lot of companies committing to zero carbon or carbon neutrality, and there's a lot of attention on it. So, we also increased our engagement. We developed the sustainability task force. We have an internal resource for that, but we're also measuring historically. For the designs we've generated so far, we commissioned an engineer to measure the carbon footprint and sustainability performance of these buildings to learn more precisely about what we need to pay attention to. I also think what is very important to me personally is not only looking at the final measure of a certain carbon footprint, which you can achieve by many things like using local labour, materials, using certain machinery or using certain recycled concrete, but what the architect can do specifically, not technical systems, but the morphology of the building and the way if you orient with respect to sun and wind, and we've done a number of projects focusing on this a lot, you can actually get a particular look and feel of these buildings. We can also look at the building and see where south, north, east, west, so it helps you to understand your environment.

It utilises natural ventilation, more passive systems. This moves self-shading, etc., brise soleil elements [a type of solar shading system that uses a series of voids and blades to control the amount of sunlight and solar heat entering a building], which change as you go from south to all directions, so they're becoming more organic, integrated, and they really also change the appearance of the building, and you can see and sense that this is a highly sustainable building, highly adaptive to climatic conditions. So, with an architectural impact, and the examples are, our central bank in Iraq [Central Bank of Iraq], the tower, which is very much based on structure and shading, structure as shading, or the Riyadh Energy Research Centre [King Abdullah Petroleum Studies and Research Centre], where we have a lot of small windows, shaded courtyards, and shading canopies, which have a similar orientation towards north and the wind. These are seminal projects for us to show that environmental and conscious architecture also generates a new appearance, and what's also beautiful is more embedded in the local architectural conditions because we're talking about passive systems, which are also operating in the pre-air conditioning era of the local architecture. There's an intuitive sense that the buildings belong to their region because they're more registering the climate and often times you find that you don't see any difference between the building, which is highly certified with sustainability, versus another because it's all hidden in some machinery or some choices, which are not genuinely architectural choices. So, that's my preference.

I also want to say that I think it's quite important to see that sustainability is a constraint. It becomes a driver in the way I'm talking about, what I call environmental tectonism. But we also need to worry about and keep worrying about the quality of our built environment for its original purposes as an agenda.

Sometimes I feel that the sustainability agenda is the only thing one talks about because it's easy to talk about, and you can measure it, and you can say, I have these certifications, and I'm a good person and conscientious person. But sometimes there's not enough time left to talk about the qualities of the experiences, of the interactions. That needs to be the primary topic—what kind of life process you want to shape, how inviting and open a building should be, how secluded some of the spaces might want to become, and then look at that together with environmental constraints. We shouldn't have the tail wagging the dog, if you know that metaphor. When we focus on something which should be a constraint, an important constraint, which could also become a driver, but it's never the primary driver for an architecture or urban formation. That's my attitude towards sustainability, and I have a new book coming out called Tectonism, where I talk about how the different engineering advances, structural engineering, but also environmental engineering can shape the built environment and the buildings in a new way, which when you also sense, oh yes, this is 21st century and this is sustainable, and you understand why because it employs these passive systems, it doesn't need so much energy to begin with. I'm quite conscious to focus my attention and talk about my architecture with respect to aspects of sustainability, which are in the genuine control of architecture and not belonging to the domain of, let's say, infrastructure or engineering or material science.

W So, is that your new book?

P Yes, it's a new book. It's coming out in September this year. It's actually going to be in Chinese as well as English. You can already pre-order it, I think. It's called Tectonism and Architecture for the 21st Century. The environmental aspects of tectonism are highlighted, but also, of course, we're talking about new robotic fabrication impacting as well as new structural optimisation technologies. I think they give a very beautiful input to architecture, and what is interesting is they make it look more like natural formations and organisms, more organic because in nature, you do have also this adaptive responsiveness to structural optimisation. In nature, everything is structurally optimised through evolution, similarly, the adaptation to natural environments you find and include in nature. Here's also, let's say, the vernacular traditions of the built environment. They were not consciously designed, but they had by necessity adapted through gradual learning and adaptation process to the climatic conditions and to the material resources of these regions. They're nearly like organisms because they had hundreds and maybe thousands of years to evolve. Contemporary requirements are maybe more stringent as we have to build larger, so it's not a literal translation. But I do find it interesting that our buildings look more like they're coming from nature and maybe also connecting somehow with the intricacy and irregularity or rule-based variation you find in vernacular structures.

W   This explanation helps me to have a deeper understanding about the sustainability, which is very inspiring. Also it's good to know there's something to expect in autumn—your new book is coming out. In the upcoming exhibition 'The New World' at Guardian Art Centre in Beijing, a selection of ZHA's international and Chinese projects, including over 120 models and many video projections, will be displayed. The exhibition will finish with a metaverse tunnel. You once mentioned that, 'The metaverse I want to contribute to supports and becomes part of productive societal life and an integral part of social production and societal reproduction. This serious type of metaverse does not offer an alternate reality or second life or any escape from social reality or societal life, but instead enhances society and enables fulfilling, productive lives.' Could you please talk about the new possibilities and opportunities brought by metaverse?

P   Thank you. You found the perfect quote. It's a good premise to clarify because a lot of metaverse projects currently are more the video game—play to earn, addressing teenagers escaping from reality and their irresponsibility. So I don't want much to do with this. What we're doing is we're having a technology transfer from these massive multiplayer environments, virtual environments, where you can interact and so on and so forth. There's a technology transfer, but really what we do with this technology is directly connecting up with our everyday lives, productive lives in particular, cultural lives, educational lives, so I think you picked the right quote, and we're working on a metaverse now, particularly addressing and inviting the people we know best, our colleagues, other architects, designers, interior designers, fashion designers, car designers, product designers, graphic designers, metaverse designers as well into a new community, a space where you can set up your shop, you can display what you're doing, you can actually come in there, have meetings, discussions. We're actually hosting at the moment, the virtual Venice Biennale in our space. It's just the beginning. We have a major building and a few other buildings, but we want to build that into a whole little design district, virtual design district, maybe even a design city. That's the idea. I think, yes, for us, it's clear that this competes with and substitutes and maybe also connects up with our work and cultural spaces and the physical world. There will be windows into these virtual worlds. These virtual worlds will also be designed similarly. Sometimes you might have clients where we design both the physical venue and the metaverse version or venue and how they connect. So it's a continuous social reality. I think it's a very important message, and that architects realise our core competency allows us, we're in charge of the user experience and not of what's happening technically under the hood. All of the physical engineering disappears, the design ideas can translate one-to-one between metaverse and physical. In the metaverse, of course, there's a totally new engineering stack and new set of constraints and requirements, polygon count and refreshment rates, and many other constraints, so the design is seamless. We have to say, if you're an architect, and you go into the designing metaverse spaces, you're still an architect. You're not leaving your profession. You're continuing your profession because it's also the same criteria of functionality you require. Is it easy to orient in that space? Can you find, can we find each other in the space? Can we participate in a nice lecture or in an intimate chat or can we visit an exhibition, or can we discuss a design maybe with a 3D model in front of us? All the criteria of what makes a successful space are the same, whether it's a physical or virtual space because we want to have similar, we achieve similar things. We want to achieve the aim, the goal of the meeting. We want to achieve the goal of the exhibition. We want to achieve the goal of the, maybe a social event where we want to get to know a few new people, but also get to know them a little bit more in depth. The requirements of life are the same, and they have now two options how to fulfill them, and that was already the case with social media, and this is already the case with things like Zoom and so on. But I feel that the metaverse, I believe in it because I think the experience is more compelling than for instance other remote alternatives.

Writing emails or chatting on WeChat is not the same as joining a virtual meeting and roaming around in a virtual city and getting to know some people. I just gave a lecture in Beijing to a Chinese audience, and there's this grid full of names and some faces, you don't know how many because then there might be several other pages, and they become very small, and you can't interact, you can't pull aside and chit-chat or have before or after the event, have some chat and informal conversation. All of that doesn't happen in a tool like Zoom, but it would happen if you did this similar lecture in the metaverse, you can have all of that. You come a bit earlier, see who's there, you know, do your lecture, let's chat, let's look at it, it's more interactive, etc. You can also see what else is going on in the building. So maybe you came from the lecture but here you can see, 'oh what's going on there,' somebody is explaining this model, and let me have a quick look and join that. So you see more, you can navigate and browse. On the internet you can't do that, you don't see other events when you're on Zoom, it's a one event by event, and in the metaverse, it's like in the real world, you go to university, you focus on something, but at the periphery of your eyes, you can sense a number of other things going on, and you can kind of switch over, etc. I think the advantages are compelling, and you can integrate that better with a real meeting. You could be several people meeting in a space with a big screen, and several others joining in, instead of you always by yourself on your desk at home. It could be socially, physically social and virtually social at the same time. These advantages, they mean to me that everybody, every organisation, everybody who has a website, everybody who has a building or space, corporation, charities, university, media outlet, everybody will have virtual spaces in virtual cities. I know that because I think the experience is compelling, and the question is when does it happen? When there's only a few things there, then it's not worth to go to the metaverse. So who's coming first? There needs to be these early adopters. It was like the internet. I remember when the internet first came out, it wasn't interesting, and then you went there, there weren't many websites, and when you tried to use it, before Google existed, you wanted to search something, it wasn't what you were looking for. So then you say, okay, no, and then something happens when things gain in quality, but also when there's more accumulated, it reaches the threshold, and then network effects kick in. Suddenly more and more get in and now in a flash, everybody has to get it. I know this moment is coming. We're preparing for this, and we will be, there will be an avalanche of work for all of us to design all these spaces. That's why we need to gear up with AI and generative processes to do them all, and I think that's the key. I'm excited this will come, but you don't know when because right now there's nothing to go to. Maybe in three years, everybody will be going. But this is such a positive vision to accept.

W Great. And now this is my last question. Let's go back to the future. In your opinion, architects need to design for the future knowledge-based and creative society, and design for interactive spaces that can stimulate imagination. You paid attention to how technology companies build their spaces, their choice of locations, and what kind of atmosphere they expect. This is also the research you have been doing with students from AA and GSD. According to your explanation, in doing so, 'civilisation will evolve accordingly, and every member of the society will have a better life.' What is your vision about a better life for people in the future from the perspective of an architect?

P I believe so. I think technology is amazing. Robotics, software, service, 3D printing, and that means production, physical. If you go to agriculture, there's very few people. You have robotic tractors and seeding, and you have intensive farming with 25 harvests in an underground with LED light. All of this is developed in cities, in tech hubs, agro-science, etc. Technology development, innovation, innovation, innovation. I think everybody will participate in this mostly, and I think the nice thing about it is, if we do it right, prosperity gains are so much that we work less or only if we stimulate it and love to work. We continue to learn at the same time. Our work is more self-directed. Nobody can control us to do all this creative work. We have to be self-initiated, and when we're tired, we can go to sleep, and when we have something else to do, we can manage our own time. At the same time, we can stay in touch. This is the better life. The better life is the more productive life and less constrained life, the free life. I think we as human beings, we don't want to be lazy or we get bored when we have nothing to do, so that's why I think this life is going to be the better life because I think we're happy when we're productive, and when we're productive in a way which isn't super stressful and we don't have somebody else telling us what to do. That's the vision I have, and for that, we need very open and easy navigate spaces. We need to use the metaverse. Even in a company like ours, we already feel that way and you can see the difference between a more advanced economy and a slightly less advanced economy that you have less command and control, more participation and self-motivated work. I think work is important to feel happy if it's not drudgery, if it's stimulating. So that's my vision, and that's also my final word on this interview.

W Thank you so much for showing us the trajectory of the Zaha Hadid Architects. We've been talking about sustainability, research and China. And we're looking forward to your new book, your new thinking, and your new buildings.

P Thank you very much. So nice to meet you and see you again.

采访大桥谕

扎哈·哈迪德建筑事务所 总监

W 大桥谕，您好。很高兴在庆祝扎哈·哈迪德建筑事务所（ZHA）在中国取得的15年成就之际采访您。您来中国已经很久了，请问您如何看待这个时期？您对中国建筑领域和整体环境的变化有何感受？能谈谈这些变化对扎哈·哈迪德建筑事务所北京办公室建筑实践的影响吗？

S 首先，我要感谢，也很高兴有这个机会讨论并庆祝扎哈·哈迪德建筑事务所在中国15年的实践，以及我们过去几年的工作。在中国，我们见证了在规模、速度层面都前所未有、不可思议的变化，我认为，这是在世界上其他任何地方都无法发生的。我们很高兴能够成为这个飞速发展的时期的一部分，并继续参与其中。如你所知，我们一直在中国开展多样的项目，从小规模的家具设计，到室内、建筑设计，从文化建筑设计，再到大规模的城市规划。我们事务所一直秉承的乐观主义、热情和勇气在这里得以实现，我们不断创新、迸发新的灵感，并将设计和创造力融入这里的整体环境。我认为这就是我们的工作最令人兴奋的地方，这就是我们为什么会成为建筑师和设计师。我们永远充满激情。

W 您在中国完成了许多重要的建筑项目。能否请您谈谈印象最深的项目并阐述其背后的原因？

S 我们所做的每个项目都是挑战，因为我们会对客户的需求做出符合环境条件、绿色可持续标准以及城市氛围的响应。每一个项目的灵感都来自自然环境，尤其是场地条件。位于北京西南部的丰台区的丽泽SOHO是近几年一个非常激动人心的项目，这座建筑就是根据场地条件进行设计构思的典型。地铁14号线和16号线的隧道相交穿过这个建筑物的下方，创造了非常独特的场地条件。我们要设计一座200米高的办公大楼，但它必须横跨位于地基正下方的地铁隧道。因此，我们设计出了一座分立为两部分的优雅、挺拔的建筑——它跨过隧道，轻轻地向上扭转，逐渐与北京中轴线的方位对齐，仿佛是回望这座城市的一扇城市之窗。它成了这个城区一座引人注目的地标建筑。作为一栋办公大楼，它的底层直接与地铁系统相连接，方便人们搭乘地铁来这里办公或使用低层的零售空间和公共广场，这座建筑在邀请和期待人们的到来。作为一个地标建筑，它也对更广阔的城市区域做出了回应。从技术角度来说，我们在设计的过程中使用了一些自主研发的新工具，如3D工具，它可以帮助我们更加精确地控制几何形状。我们不仅可以控制建筑的设计，还可以控制建筑物、幕墙系统的制造与安装。这项技术还能使建筑的设计更加融入自然环境，例如，使自然光和风进入建筑物。在建筑中，你不仅可以欣赏城市的全景，还可以享受自然采光。建筑白天消耗的能量更少，从而提高能源效率。同时，我们还设置了一个能过滤、净化空气的自然通风系统，因此，建筑物内部的空气会比建筑物外部的空气更加洁净、清新。所有的这些都整合在一座非常优雅的建筑中，我们对此感到非常兴奋。我认为它已经成为城市的一部分了。

另一个让我们印象深刻的当然是北京大兴国际机场，这是一个非常具有挑战性的项目。作为建筑师，我们要考虑的是如何打造一个符合北京，乃至整个中国规模的机场和新门户——每年需要接待超过7200万，甚至超过1亿的乘客。我们要考虑的是如何打造世界上最大的机场之一。因此，我们在这个项目里再次回应北京城的南北中轴线，将中轴线延伸至大兴南部地区。这创造了一个交通枢纽——高铁及所有铁路系统、公共交通和私家车辆等都可以高效地直接到达机场。一旦抵达机场，旅客就会进入一个非常实用且灵活的空间。我们设计的这个星形机场的整体理念以效率为核心，考虑了接收人流、车流、办理登机手续等程序的效率。只需几分钟，旅客就能到达机场的中心，并且可以在8分钟内走到最远的登机口。也就是说，旅客通过8分钟即可步行感受这座美丽的建筑，同时行至最远的登机口。所有的这些都是在项目概念的早期阶段——概念图设计阶段就计划好的，并且是由一个非常大的团队来完成的。这不仅包括我们的团队，还有来自世界各地其他团队的紧密协作。例如，中国本土的团队——北京市建筑设计研究院，以及法国巴黎机场工程公司（ADPI）。所有团队齐心协力，共同打造了一个能够真正代表北京甚至中国的非常激动人心的项目。我们以非常快的速度完成了这个令人难以置信的项目，这也是非常具有挑战性的——这个机场仅仅用时5年就建成了。我在想，世界上没有任何其他地方可以在这么短的时间内建成如此庞大的基础设施。我们能够有幸参与其中，真是让人难以置信。

W 感谢您与我分享这些很有意思的项目。现在，我们来聊聊您的团队吧。目前，您在中国有25个项目正在进行中。能否请您谈谈您是如何获得设计灵感，如何为北京的团队创造这样一个能够不断孵化新创意的环境的？

S 我们非常幸运，2008年就能够创建ZHA北京办公室。多年来，我们的工作室稳步成长，我们不仅能够交付、真正实现我们所设计的项目，而且能够将新的方法、新的技术融合

汪芸→Wangyun→**W**
大桥谕→Satoshi Ohashi→**S**

到设计过程中。当然，我们的北京团队和伦敦团队都是这种创造精神不可或缺的部分。我们将事务所承接的每一个项目都视为全新的项目，会针对每个项目的不同条件进行用户化的定制设计。我们会根据地点、客户以及环境，挑战自己。我们从不重复，总是对新的可能性满怀期待。这种精神一直鼓舞着我们。正如我刚刚所说的，我们常通过建筑所处的环境、场地条件获得启发，因此我们常常有机会探索不同的可能性。我们也开发出了越来越多新的工具和技术，因此，我们对事物之间的联系有了更多了解，比如，理解阳光的轨迹，理解水和风的运动。我们在受到大自然启发的同时，也理解人的流动、城市的流动，甚至资金的流动，以及世界各地正在发生的变化。这也会对我们的设计工作产生启发性的影响。我们的北京办公室是由一群非常有才华、充满活力的人才组成的，我们因为相信一切皆有可能而相聚于此。我认为正是因为这种精神——我刚刚谈到的我们事务所的DNA，我们才能在中国取得成功，并继续在世界范围内发展。

W　现在我们了解了您的团队一直如此富有创造力的原因。让我们来看下一个问题。您曾提到："我们葆有对设计品质永无止境的追求，初心不改；事实上，新设计工具和通信技术的整合将这种精神进一步深化。"请问随着技术的发展，您如何调整、优化设计质量的标准？

S　我们正处于一个激动人心的时代，这确实是一个充满活力和变革的时期，不断变化的新技术对我们的生活、工作以及享受生活的方式产生了不可思议的影响。现在，我们每天都能听到这样的说法：新的AI技术将改变我们生活方式的方方面面，人类发展将面临下一种可能性。我认为我们必须对此保持高度关注，同时为新技术的整合和利用问题担负起责任。我们有责任利用这些技术造福人类，为所有人创造更好的区域和空间。我认为，（新技术）将对我们的设计方式和设计目的产生影响。通过利用新技术，我们能够持续学习并不断发展。我们一直在实验、探索。实验，一直是我们办公室研究与开发环节的重要部分，我们不断将其整合到整个设计工作的流程中，因此能在开发设计工作中保持领先、前沿的精神。

W　太精彩了。我们知道，除了出色的建筑实践之外，您还是一位优秀的老师。您曾在世界各地的多所大学任教和演讲。目前，您正担任中央美术学院和清华大学的客座教师。请问，您认为您和ZHA给中国年轻建筑师带来的最重要的东西是什么？

S　我有幸参与了世界各地许多大学的教学工作。我曾受邀在中国的几所大学担任访问评论员。教学工作最吸引人的地方在于，你会看到充满好奇心的年轻一代，他们见证了ZHA的设计师如何吸收想法和概念，并在建筑和新的城市规划策略中真正开发并实现这些项目。这真的正在成为现实。正如我之前提到的，我们有许多项目实际上正在建设中，很快就会成为现实。这些项目不仅在概念层面上得到了发展，而且在转化为现实的整个过程中也得到了进一步的开拓。我们了解如何开发新技术和交流的工具，我们会将这些知识传授给年轻的人才，或者将这些知识进一步传授给实际着手共同开展项目的制造公司；之后，再普及到更远的下线，直到在建筑中得以实现。项目落成，才能真正回馈给年轻一代，对他们产生影响和激励。我认为这是一种积极乐观的态度，这种怀揣梦想并对未来充满愿景的状态对于年轻建筑师或年轻一代至关重要。

W　说到这里，我有一个问题。请问您希望在这些年轻的人才身上培养哪些特质？您强调梦想、灵感和激情，请问您是否有特定的方法来激励他们发展这种品质？

S　是的，在每个项目里，我们都希望能够实现自我挑战、自我突破。我认为我们一直在努力工作着，我们一直保有需要努力实现的目标和梦想。年轻人需要有协作精神，需要聚在一起并共同努力实现愿景，共同解决问题，共同创造这样的环境。这非常关键，不仅在教育层面，在实践层面也是如此。在ZHA，我们所有人都聚在一起，有着挑战自己的实践精神，永无止境的学习精神。因此，我们能为这个世界创造出非常新颖且更美好的事物。我认为，培育这种精神并且保持这种精神才是最大的挑战，只有这样，人们才会愿意付出努力、持续为梦想和未来而拼搏。这不仅是教育的重要环节，更是我们解决这个世界未来所面临的难题的方法。只有保持乐观的态度，我们才能为人类创造更加美好的生活、工作和享受生活的空间。这非常振奋人心。

W　这确实非常鼓舞人心，谢谢。您刚刚探讨了灵感与梦想。现在，我们来谈谈您对未来的愿景。您曾经提到科技、计算机运转速度的进步和工具的发展影响着人们的思维方式和生活方式，我们处在一个不断变化的环境中。那么，ZHA北京办公室未来15年的愿景是什么？您认为ZHA北京办公室将面临的最大的挑战是什么？

S　我认为，我们的办公室在不断发展。15年前，我们为了真正扎根在这里而来。我们了解如何设计，当然，还了解如何深化和真正实施这些项目，我们能够通过设计对中国的情况做出回应。这些年来，我们已经掌握了如何以中国速度和中国规模来做到这一点。在接下来的几年里，伴随着技术更快、更进一步的发展，我们需要谨慎留意这几个方面：如何理解这些工具，如何运用它们发挥有益于我们的作用，以及如何更好地用它们来开发更多对我们有帮助的工具。我们现在面临人工智能和技术的进步，这可以丰富我们做事的方式。事实上，这让我们重新审视过去所做的，并思考这到底是否有益于每个个体。我们需要对此保持谨慎，归根结底，这关系着我们对人类所负的责任以及我们究竟可以做些什么。这些技术实际上是我们社交空间的延伸，技术使我们能够更多地交流、产生联系，以及分享知识。如果这可以通过人人都适用的方式实现，那我们也将从中受益，并获得其他可能性。当然，伴随着不断的变化，我们能否跟上这些变化和发展，能否保持多年来已经融入我们DNA的前沿精神，也是巨大的挑战。因此，我们需要继续引进新的人才，并且对预见并整合新技术继续展开研究。我认为这将是我们永远的事业，也正是我们所乐于做的事情，我们希望并期待能继续这样的实践。

W 这段话对今天的采访做了一个很好的总结，您探讨了知识、沟通、不断变化的环境、前沿精神，以及反思和批判性思维。非常感谢您接受采访并给出精彩的回答。

S 很高兴有机会和你进行这次谈话。确实，我们期待能够继续这段旅程。从草创阶段到如今，这真的是所有参与其中的人共同努力的成果。我们从未改变初心，这种精神是我们的起点，包括帕特里克·舒马赫在内的事务所的创始人以及所有核心团队成员在过去10年、20年、30年中一直保持着这种精神。我认为，这就是自始至终推动我们事务所前进的原动力。我们将会一如既往地继续下去。

W 这着实鼓舞人心。非常感谢您分享了您的真知灼见与精彩实践。

S 谢谢。我们保持联系。

Interview with Satoshi Ohashi

Director, Zaha Hadid Architects

W Hello Satoshi. It is such a pleasure to interview you during the time period when you are celebrating the 15-year achievements of Zaha Hadid Architects (ZHA) in China. You have been in China for quite a long time. How do you look at this time period and experience the changes in the Chinese architecture field and the overall environment? Can you talk about the impact of these changes on the practice of ZHA Beijing office?

S First of all, I'd like to thank you all for this opportunity. It is great to have this chance to discuss and celebrate, of course, the 15, 16 years of our work in China, as well as just Zaha Hadid Architects, the work over all the past years. Of course, in China, what we have seen is the incredible change in scale, speed, and that is unprecedented in the whole world. I think nowhere else in the world can you see the scale of change that has occurred in this period of time. And we are very excited to have been or continue to be part of that period of change in China. We have been working on many, many projects here, as you know, from the small-scale product design, furniture design to, of course, interiors, buildings, and cultural buildings to large-scale urban planning of cities. We have been able to continue that unyielding optimism, that unyielding passion and spirit that has been part of the office and this creativity to continue to innovate, inspire, as well as integrate design and creativity into the whole environment. And I think this is what is always exciting about our work because really, this is why as architects and designers, this is what we do. That is fantastic.

W You have completed many important architecture projects in China. Could you please talk about the project that impressed you the most and explain the reason behind it?

S Well, every project we do is a challenge that we like to respond to all the conditions, the needs, as well as the conditions of the environment, responding to the green and sustainability issues, as well as the needs of the client, and, of course, responding to the city. Each one is always inspired by nature or the landscape and specifically those site conditions. I would say that some of the recent projects that are quite exciting is Leeza SOHO, which is located in Beijing. As you know, it is in the southwest district of Beijing called Fengtai District. It really is designed and conceived because of the site conditions. There is a service tunnel of the subway lines, line 14 and 16, that cross underneath our site, which creates a very unique condition where we are designing an office tower building 200 metres tall, but we have to straddle the subway service tunnel that sits directly underneath, in the foundation below. So, what this does is we come up with a concept that we generate a building that is not only elegant and rises up but is split in half. Then straddling the tunnel, it gently twists upward to realign itself with the grid, the direction of the main axis of Beijing. It's almost like an urban window that's looking back at the city. This gesture really creates a very iconic landmark for the new district, which not only functionally becomes an office tower but also connects at the lower levels directly to the subway system. It allows people to come to work as well as use the lower-level retail spaces directly from the subway, creating public plazas, creating spaces that can be used by the people at the lower level and ground level, inviting people into the building. Then what happens is it is also responding at the larger urban scale by gently twisting and creating an iconic landmark for the city, looking back at the city. This, of course, technologically because it allowed us to use some of the new tools that we developed, the 3D tools that really can control the shape, the geometry very accurately. We can not only control the design, but also control the making or the fabrication of the building or the façade system or the windows as they step around and create this very exciting façade. It is also responding to the environment because it allows the natural sunlight and ventilation to filter into the building—you have panoramic views of the city but also natural daylight throughout that allows it to be more energy efficient. You're using less energy during the day, and then at the same time, creating a natural ventilation system that allows a filtration system that cleans the air. The building has much cleaner air inside than outside the building. Therefore, all these things have been integrated into one very beautiful building. We are very excited about that, and I think it has become part of the city.

The other one that of course really impresses us is the opportunity to work on the new Beijing Daxing International Airport. It was a very challenging project. How do you respond to the scale of Beijing and the scale of China, representing a new gateway for the country to allow over 72 million passengers per annum to use the airport, and then to have expanding capability to over a hundred million to be one of the biggest airports in the world? So here we are responding to, again, the city, the main central axis, what they call the grand axis of Beijing from running north-south and anchoring that axis in the southern part of the Daxing area. This also creates a transport hub, which the high-speed train and all the railway systems, as well as public transportation, the buses and the vehicular traffic can all arrive very efficiently and directly to the airport. Once you arrive, you are welcomed into a space that really is very functional and flexible. The whole idea of the airport of this star-shaped plan that we designed was about efficiency—how to receive the flow of people, the flow of traffic, and you check in and within a few minutes, you are at the centre of the airport, and you are

汪芸→Wangyun→**W**
大桥谕→Satoshi Ohashi→**S**

able to then walk to the furthest gate in eight minutes. You have this beautiful eight-minute walk to the furthest gate. All this was planned at the very early stages of the concept and the schematic design phases of the project, and it was done with a very large team. Not only our team, but also the teams around the world working very closely together—the local team in China was the Beijing Institute of Architecture and Design (BIAD) and in France ADP Ingeniérie (ADPI). All these teams coming together to collaborate and create a very exciting project that really is representing Beijing as well as China. I think these have been very challenging but also incredible projects that were built very quickly. The airport was built in five years, and I think again, nowhere else in the world can you build such large infrastructure within such a short period of time. To be involved in that as an opportunity has been incredible.

W Thank you for sharing with me about these interesting projects. And now, shall we talk about your team? Because currently you have 25 projects going on in China. Could you please talk about how you get inspiration for design and to establish an environment that would constantly incubate the creativity in your team in Beijing?

S We have been very fortunate, since 2008, we have been able to establish the ZHA Beijing office. Continuously and organically growing steadily over the years, but at the same time, being able to design, deliver, and really build all these projects, integrating new methods, new technologies and integrating that in the whole design process. Of course, our team here, as well as in London, is a very integral part of that spirit of creativity. We are able to take on every project as if it's a new project that wants to be specifically customised for that condition, and depending on the location and the client, as well as the conditions, we challenge ourselves. We never try to repeat anything. We always want to see what is possible. This spirit has always allowed us to be inspired. We are also, as I said, often inspired by those surrounding conditions or landscape because they always give us an opportunity to test different possibilities, and as we have developed new tools and technologies, we understand the new relationship of things. We are able to understand the movement of the sunlight. We are able to understand the movement of water and the wind. We are often inspired by nature, but we also understand the movement of people, the flow of the city, the flow, maybe even of money, the flow of the change that is happening around the world. And this also inspires us or influences our designs when we are working. The office in Beijing is made up of a collective of very talented, energetic individuals who have come together for the purpose of all believing that things are possible. And I think it is this spirit that, as I mentioned, is the DNA of the office that continues to allow us to be successful in China, as well as continue to expand around the world.

W So now we know the reason behind why your team can continuously be so creative. Let's move to the next question. You once mentioned, 'The endless pursuit of design quality and the spirit from the early days has never changed; in fact, it has just been pushed further ahead with the integration of new design tools and communication technologies.' How do you adjust the standard of your design quality as technology evolves?

S We are facing a very exciting period, but at the same time, really a dynamic period of change, ever-changing new technologies that have an incredible impact on the way we live, work, enjoy life. Even now, every day we hear about the new AI technologies, as well as the next possibilities that will change every aspect of the way we are living, as well as humanity. I think we have to continue to be aware, as well as be responsible for how we integrate all these new technologies. And it is our responsibility to use the technologies for the good of humanity and with creating better places and spaces for all of us. I think this will then impact how we design and what we design for. We are using the technology as tools, and we are continuing to learn and to evolve. And we are always experimenting. There is an endless experimentation that has been part of, again, our office. This research and development that we continue to integrate back into the whole design workflow or the process allow us to stay ahead and keep our cutting-edge spirit, as well as our design work that we develop in the office.

W That is great. And we also know, besides your excellent architectural practice, you are also a wonderful teacher. You have taught and lectured at many universities around the world. Currently you are a visiting critic at Central Academy of Fine Arts and Tsinghua University. What do you think are the most important things you and ZHA are bringing to the young Chinese architects?

S Yes, I have been fortunate to be involved in many of the universities around the world and as well locally here. I have been invited to be visiting critics at the local universities here. And what's very fascinating is that you see the young generation who have curiosity and who can see how Zaha Hadid Architects has been able to take ideas and concepts and really develop and make real these projects in architecture and new urban planning strategies, that are really becoming real. As I mentioned earlier, we have many projects that are under construction and will become reality very soon. These projects have been developed not just at a concept level, but also the whole process of how we can make it real. We have an understanding of how to develop the technology, develop the tools to communicate, and then also transfer that knowledge to the young talent or again transfer that knowledge further to the actual fabrication manufacturing companies that are actually making our projects together, and then transferring again that further down line and it is built on the site. Then once it becomes a reality, it really goes back and feeds back into the younger generations to influence and inspire them, and I think this is something that's a positive optimism that's very important for young architects or the young generation to have dreams and also be inspired for the future.

W Is there any specific qualities you want to cultivate from these young talents? Or I would say because you emphasised about the dream, the inspiration, and the passion, but do you have a specific way to stimulate them, to encourage them to develop this kind of quality? Can you explain?

S Yes, I think with every project we like to also push, continually push and challenge ourselves. I think we are always working hard. We have a goal or a dream that we need to achieve and work towards. I think you need to have a collaborative spirit and bring people together and work together to have a vision, but also to solve these problems or to create that environment or nurture that environment. I think this is very critical, not only at the education level, but also in practice. At ZHA, we all come together to have that spirit to continually challenge and test ourselves, to have that endless spirit to learn as well as to take on the challenge so that we can crevate something very new or better for the world. I think the biggest challenge is to continue to sustain that talent as well as to nurture and foster that spirit so that people always want to give that effort and continue to work towards the dream and the future. I think this is an important part of not only education, but also how the world is going to solve the future problems that we face and that optimism for the future so that we can create a better live, work, enjoy life place for everyone. That's very, very inspiring.

W That is very inspiring. Thank you. You were talking about inspiration and the dream. Now let's talk about your vision about the future. You once mentioned that the technology, the advancement of the speed of computer and the tools impact the way people think and live. We are in a constantly changing environment. Then what is the vision for the ZHA Beijing office for the next 15 years, and what do you think might be the biggest challenge?

S I think that our office continues to evolve. In the last 15 years, we came to China to really be rooted here. I think we understand how to design, of course, develop and really implement these projects to respond specifically for China. I think we have learned how to do this and how to do that in the speed and scale that is required for China. I think in the coming years, as technology advances even faster and further, we need to be careful on how we understand these tools, how we can use them again to our advantage, and how we can use them better to develop further tools that help us. I think the advancements in AI and technologies now we face, it really is another tool that allows us to expand the way we do things. It allows us to rethink what has been done in the past, and is it really better for everyone? We need to be careful about this because in the end, it's about our responsibility to humanity and what we can do. These technologies are really an extension of our social space, and they allow us to communicate and connect more as well as share knowledge, if this can be used in the way that will benefit all of us as well as provide us other possibilities. The biggest challenge, of course, is as we continue to change, we also need to keep up, change and continue to develop and allow us to keep that cutting-edge spirit that has been part of our DNA and our office for all these years. That means we need to continue to bring new talent as well as be able to foresee and integrate new technologies and continue that research. I think this is something that we will always continue to do so that we can always keep that exciting spirit of what is expected of us. I think this is what we love to do, and we hope we can continue and we are looking forward to doing that.

W It is such a wonderful conclusion for the interview today because you were talking about knowledge, communication, the changing environment, the cutting-edge spirit and the reflective and critical thinking. Thank you very much for the interview and for your wonderful answers.

S It is a pleasure to have the opportunity to have this talk with you. We are looking forward to continuing this journey, and it is really the efforts of all the people that have been involved throughout the years from the very, very early years to now. Nothing has changed that spirit. I think this is what started with, of course, founders, Patrick Schumacher and all the core group of the office that have been with the office working 10, 20, 30 years together. I think this is what always pushes our office. And I think we would like to continue that.

W It is truly fantastic. Thank you very much for sharing your insight, thoughts and intelligence, your practice. And I wish you have a great day.

S OK, thank you. Yeah, let's keep in touch any time.

展 览

Exhibition

"未来之境"展览介绍

扎哈·哈迪德建筑事务所

> 技术的快速发展和我们不断变化的生活方式为建筑学提供了一个全新的、令人振奋的背景，在这个新世界的背景下，我认为我们必须重新研究那些流产和未经检验的现代主义实验——并非为了使之复活，而是为了揭示新理念和创造。
>
> ——1983年，扎哈·哈迪德

在这个新世界的语境中，扎哈·哈迪德建筑事务所（ZHA）的作品回应了当下的挑战。当代城市的生活方式逐渐发展得复杂而多样，因此产生了需要各种建筑类型的多样的、相互重叠的社区。这种不断增强的合作性社会融合、交流和知识互换将21世纪的生活方式与早期倾向于分离和重复的现代建筑区分开来。

于2010年竣工的广州大剧院是ZHA在中国完成的第一座建筑。此后，又有14个开创性的项目相继落成，包括北京丽泽SOHO和大兴国际机场。此次展览重温了这些具有重要意义的项目，并展示了正在各地建设的25个项目。

展望未来，此次展览重新思考了ZHA对设计的深入探索和多学科研究方法。新的数字设计工具、机器人技术、3D打印技术、人工智能和虚拟现实正在改变ZHA的设计和建造方式，提升建造技术并提高材料效率，在显著减少建筑对环境影响的同时，优化每座建筑的性能。

ZHA致力于建筑行业持续迭代的研究和开发，包括有助于在每个项目的整个生命周期中减少碳排放的新的环境分析工具。ZHA与国际领先科研机构合作的可持续策略和材料的研发成果广受赞誉。此外，ZHA正在应用高级数据分析和多智能体系统，使建筑能够实现个人和群体福祉。

展览还展示了ZHA对信息丰富且社会参与度高的元宇宙虚拟建筑的探索。ZHA的虚拟环境凭借照片级的3D多人在线视频游戏创作工具，结合高速网络和云技术，专注于用户体验和对社交互动的促进。ZHA的建筑在参数化设计技术支持下，结合群体互动形成社交基础设施，正在增强与世界各地的交流与协作。

此次展览阐释了ZHA对未来坚定不移的乐观态度以及对于建筑设计一贯坚持的创新性。

Introduction to 'The New World' Exhibition

Zaha Hadid Architects

Technology's rapid development and our ever-changing lifestyles have created a fundamentally new and exhilarating backdrop for architecture, and in this new world context I felt we must reinvestigate the aborted and untested experiments of Modernism—not to resurrect them, but to unveil new ideas and innovations.

Zaha Hadid, 1983

Within this new world context, Zaha Hadid Architects' (ZHA) work responds to the challenges of contemporary urban life becoming ever-more complex with its many diverse, overlapping communities requiring a wide variety of differing architectural typologies. This new intensity of cooperative social integration, communication and knowledge-exchange differentiates our lives in the 21st century from earlier periods of modern architecture in which buildings favoured separation and repetition.

Completed in 2010, the Guangzhou Opera House was ZHA's first building in China. Another 14 pioneering Chinese projects have been opened since, including Beijing's renowned Leeza SOHO tower and Daxing International Airport. The exhibition revisits these seminal projects, together with the further 25 projects across the country currently in development by ZHA.

Looking to the future, the exhibition examines ZHA's detailed research and multidisciplinary approach to their work. New digital design tools, robotics, 3D printing, artificial intelligence, and virtual reality are changing how ZHA designs and constructs; developing efficiencies in construction techniques and materials that significantly reduce environmental impact of their architecture while enhancing the performance of each building.

Investing in the research and development of continual improvements for the construction industry, including new environmental analysis tools that contribute to the reduction of net carbon emissions throughout the life-cycle of each project, ZHA collaborates with leading scientific institutions around the world to develop award-winning innovations in sustainability strategies and materials. Additionally, ZHA is now applying advanced data analytics and agent-based modelling, enabling their architecture to address both individual and overall wellness.

The exhibition also explores ZHA's new virtual architecture within the metaverse that is information-rich and socially engaging. Using photo-quality, 3D and multiplayer online video-game creation tools combined with high-speed network and cloud technologies, ZHA's virtual environments focus on user experience and interaction. Powered by parametric design technology and incorporating community-forming social infrastructure, ZHA's architecture is enhancing communication and collaboration around the world.

Showcasing the firm's unwavering optimism for the future, the exhibition presents the consistently inventive nature of ZHA's architecture.

p34—p45,《未来之境》,数字绘画
p34–p45, *The New World*, Digital Painting, 2023

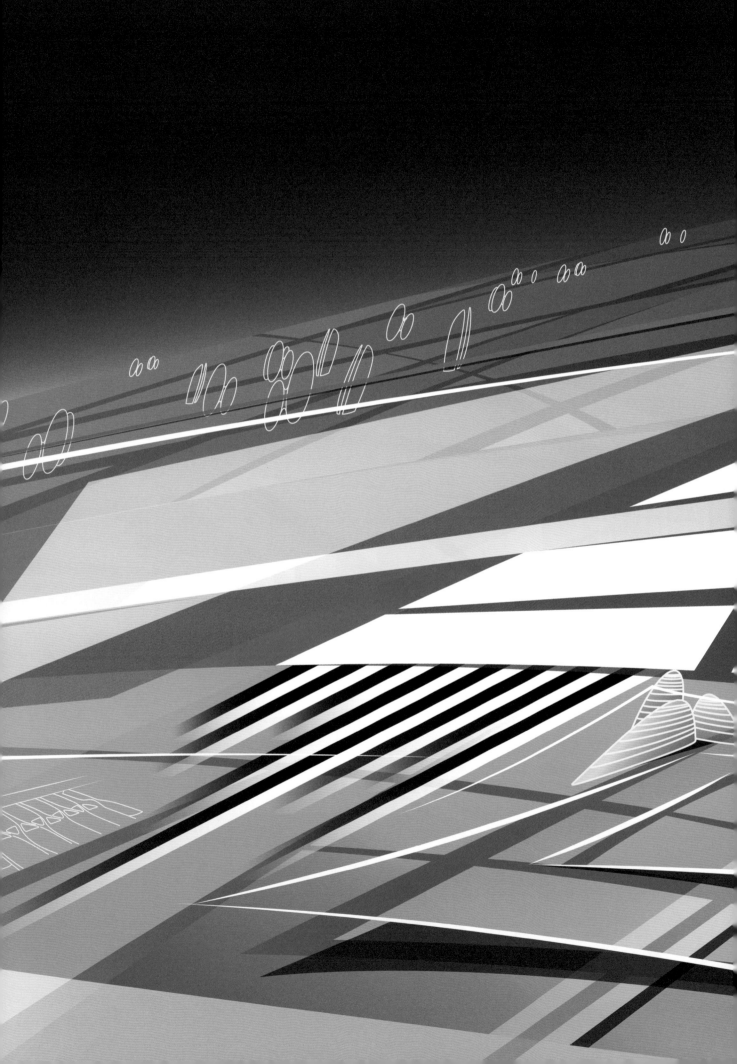

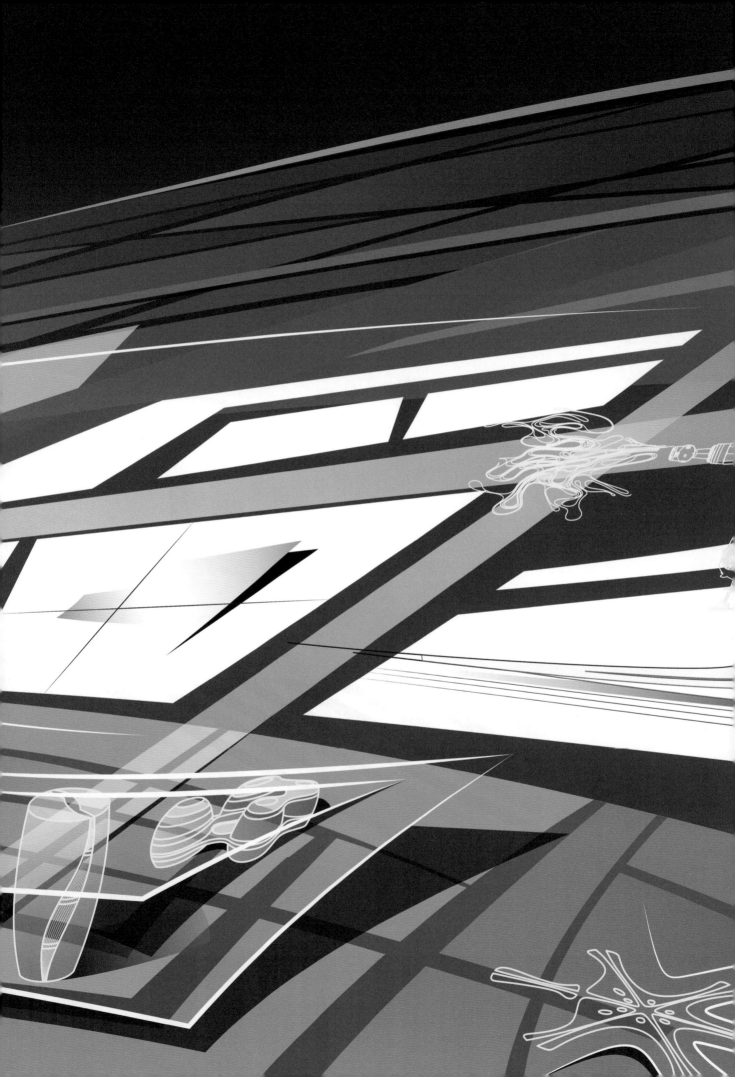

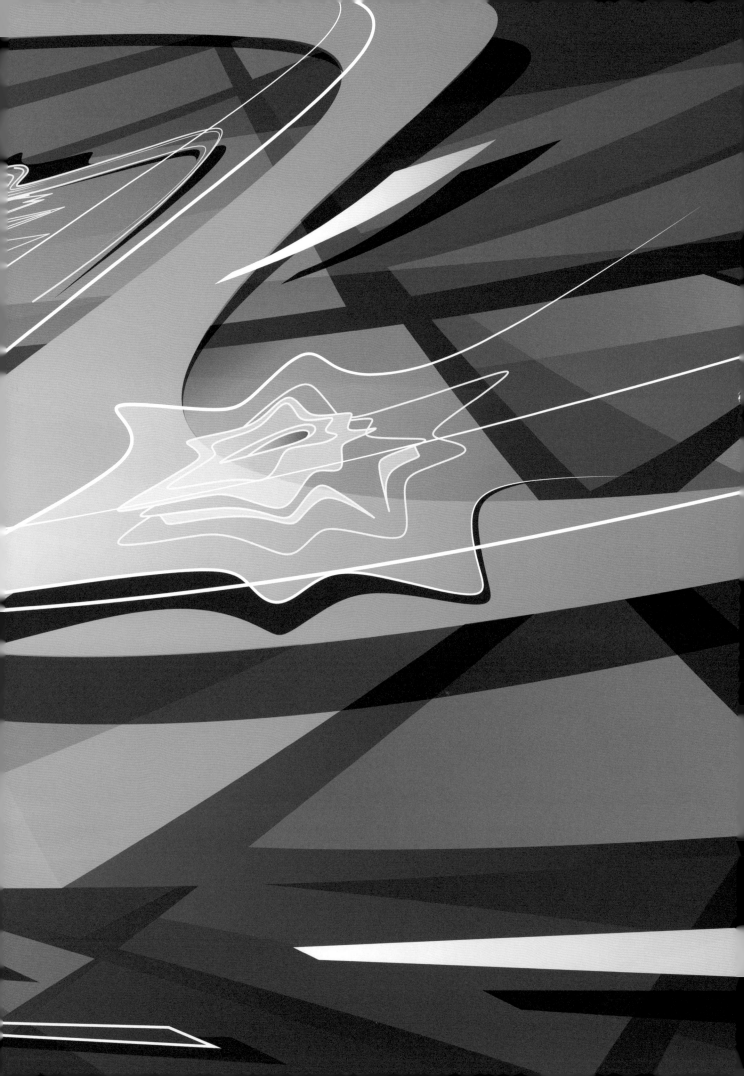

p1—p16. "未来之境"展览搭建现场照片,摄于2023年6月
p1–p16. Construction photographs of *The New World* exhibition, June 2023

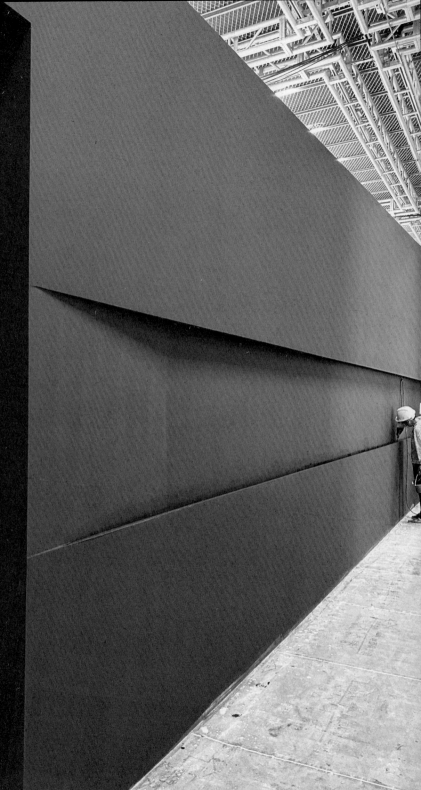

Zaha Hadid Architects
THE NEW WORLD
未来之境
扎哈·哈迪德建筑事务所设计展

p46—p77，"未来之境"展览现场照片，摄于2023年
p46–p77, Photographs of *The New World* exhibition, 2023

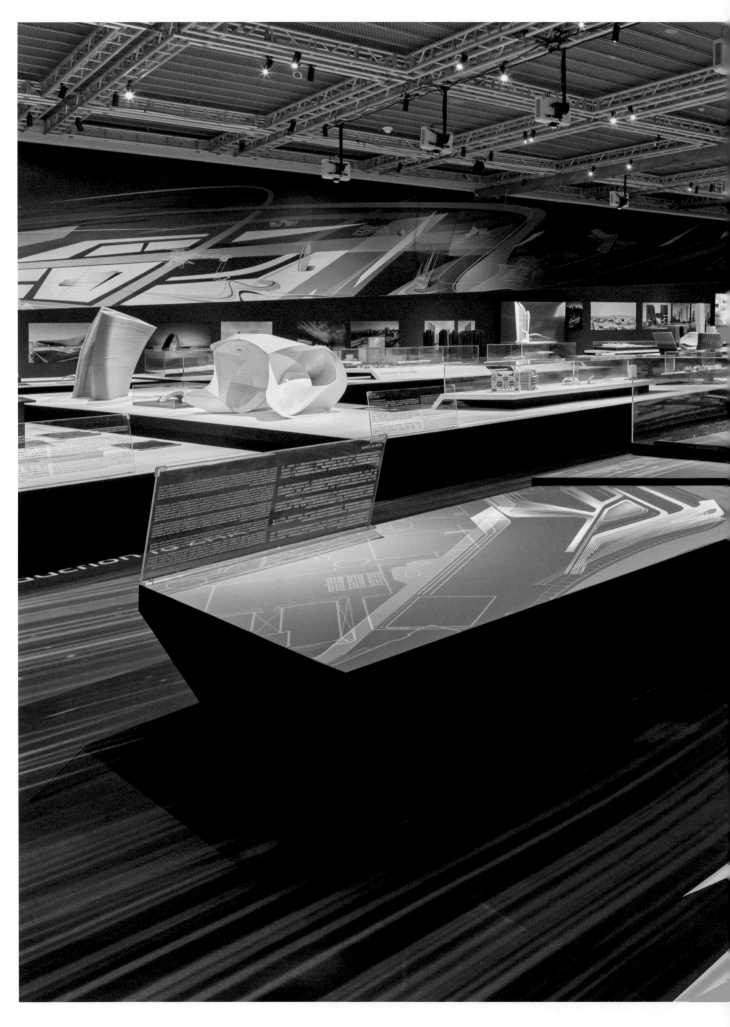

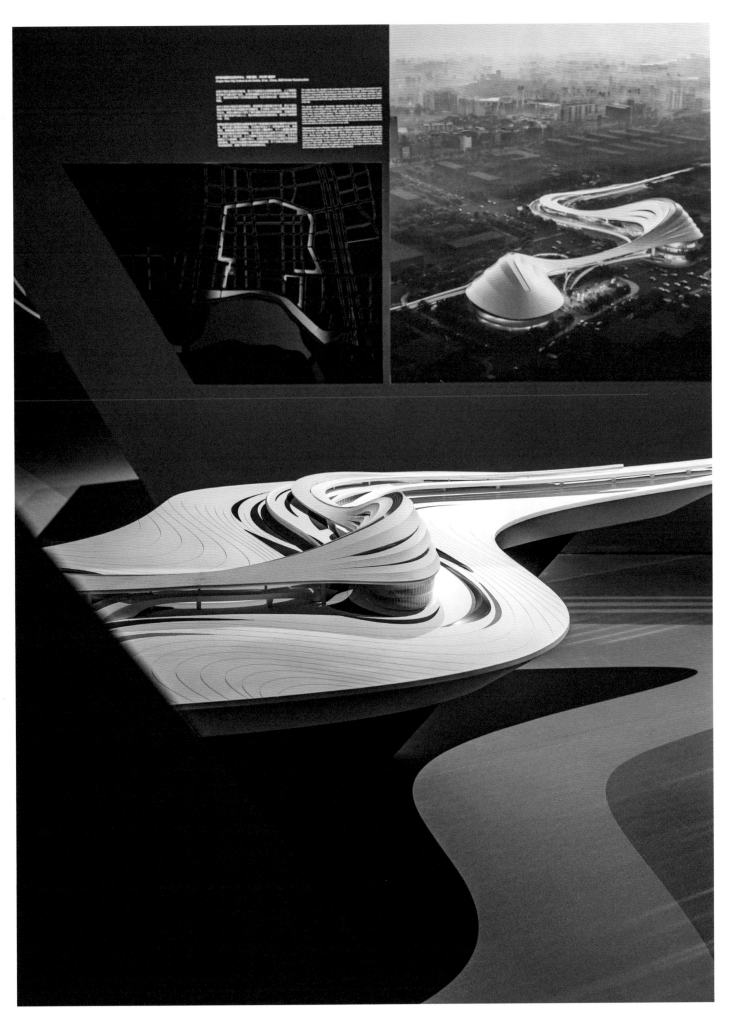

扎哈·哈迪德建筑事务所序章

扎哈·哈迪德建筑事务所（ZHA）通过诸多前卫且独具一格的作品重新定义了21世纪的建筑，作品中融合了世界各地的文化与ZHA惊人的想象力。事务所以充满活力和创新的作品而闻名于世，这些作品无一不体现着扎哈·哈迪德女爵士在都市主义、建筑和设计等相关领域的革命性探索与研究。

ZHA在全球获得了诸多民间、专业和学术机构的最高荣誉，40年来一直走在建筑、艺术和设计的前沿。这40年的创新研究全都体现在本次展出的70多个享有盛誉的项目中，为六大洲的业主建造品质卓越且面向未来的城市空间，例如，意大利罗马的"MAXXI：二十一世纪美术馆"和阿塞拜疆巴库的阿利耶夫中心。ZHA是推进建造方式不断进步的行业领袖，目前在全球范围内拥有80多个正逐步实现的先锋设计作品。ZHA将先进的设计技术与生态材料和可持续建筑实践相结合，着眼于将不同专业作为一个整体来理解，以回应各个时代的愿景。

如今，ZHA是一家拥有400余名讲着44种语言的员工的国际公司，在伦敦、北京、香港和深圳设有办公室，在全球50多个国家或地区书写着各地的城市风貌，建造代表最前沿的设计与施工技术的项目。事务所的一些作品被纽约现代艺术博物馆和洛杉矶盖蒂博物馆等许多博物馆永久收藏。超过40名员工积极投身于世界著名的设计与建筑学府的教育事业，如建筑联盟学院和伦敦大学学院的巴特莱特建筑学院。ZHA在研发方面持续广泛地投入，坚信知识共享和协作的变革力量，并致力于促进社会的积极变化。

Introduction to Zaha Hadid Architects ①

Zaha Hadid Architects (ZHA) redefines architecture with a repertoire of projects that have captured imaginations across the globe. The practice is known internationally for dynamic and innovative projects which build on Dame Zaha Hadid's revolutionary exploration and research in the interrelated fields of urbanism, architecture, and design.

Receiving the highest honours from civic, professional, and academic institutions worldwide, ZHA has been at the forefront of architecture, art, and design for four decades. These 40 years of innovative research are inscribed within more than 70 award-winning projects, such as MAXXI: the Museum of XXI Century Arts in Rome, Italy and Heydar Aliyev Centre in Baku, Azerbaijan on show in the exhibition, built for clients with reputations for excellence and visionary communities across six continents. With over 80 pioneering designs currently in development around the world, ZHA is a global leader in improving how the industry constructs. Marrying advancements in design technologies with ecologically sound materials and sustainable construction practices, ZHA does not look at the disparate parts, but works to understand them as a whole to meet the aspirations of each new generation.

Today, ZHA is a global company of 400 staff speaking 44 languages with offices in London, Beijing, Hong Kong, and Shenzhen, forming the face of cities and working on the cutting edge of design and construction in more than 50 countries around the globe. Works of the office form part of the permanent collections of Museums, such as the MoMa in New York and the Getty Collection in Los Angeles. More than 40 ZHA staff teach actively in some of the world's most recognised design and architecture schools, such as the Architectural Association and the Bartlett School of Architecture at UCL in London. Investing significantly in research and development, ZHA is driven by the belief in the transformative power of knowledge-sharing and collaboration and is committed to catalyse positive changes into society.

富谷办公楼

日本，东京
1986年
K-One Corporation

富谷办公楼对于热闹非凡的东京市区与街道来说，是与众不同的存在。透明的玻璃亭台被抬升，打造出一个微型城市空间，使周边高密度社区获得一丝喘息的机会。这是一个悬在两个水平面之间的小而紧凑的空间。该项目的绝大部分空间下沉到地面层以下，从两侧边缘向内凹，通道采用高大的玻璃幕墙，使自然光可以渗透到较低的区域。入口阶梯延伸至中层平台和下层空间的外部庭院，为位于地下的中层空间提供了向上的视野。

宽敞的下层空间使其拥有高度的灵活性，无论作为单一连续的楼层还是一系列平台，都可以满足各种商业活动和办公需求。抬升的亭台虽是完全独立的，但也是整体建筑概念中的一个关键部分。这是一个轻型单层结构，优雅地盘旋在这片空间之上。曲线屋顶和三面落地窗是该结构的亮点。这一设计为办公或商业空间提供了多种可能性。地面层的特定区域采用半透明玻璃铺装，可在夜晚照亮道路并在日间过滤阳光，同时为地下空间提供柔和的光线。

Tomigaya Building

Tokyo, Japan
1986
K-One Corporation

The relationship of the Tomigaya Building to the site and the street is a unique one in the bustling city of Tokyo. A transparent glass pavilion is elevated to create a small urban void, providing respite from the density of the surrounding neighbourhood. This void is a compacted space suspended between two horizontal planes. A significant portion of the programme is sunk below the curving ground floor, which is recessed from the edges on two sides and supports a tall glass wall in a channel, allowing natural light to permeate the lower area. The entrance stairway descends to a mid-level platform and an external court within the lower space, offering views upwards past the ground level to another platform and the belly of the pavilion above.

The generous proportions of the lower space allow for complete flexibility, whether as a single continuous floor or a series of platforms, accommodating various retail and office activities. The elevated pavilion, although completely independent, is an integral part of the overall building concept. It is a lightweight, single-storey structure that gracefully hovers above the open ground, characterised by a curving roof line and full-height windows on three sides. The design provides versatility for office studios or retail spaces. Certain areas of the ground feature translucent glass paving, illuminating pathways in the evening and filtering daylight, while creating a gentle and defuse light for the space below.

麻布十番办公楼

日本,东京
1987年
K-One Corporation

在东京,大多数场地都因为可用空间有限而受到严格限制,许多建筑只为解决人口密集带来的城市环境问题而存在。因此,在这个项目中,ZHA提出了释放东京城市空间的策略。

麻布十番办公楼位于六本木区附近的一个由各种建筑组成的峡谷之中,在景观中雕琢而成并且深入腹地,有意地强化其狭窄场地带来的压迫感。该建筑的原始玻璃结构被压缩在高耸的金属墙和钢筋混凝土墙之间。办公楼有两面幕墙——一面是醒目的蓝色玻璃幕墙,另一面是向外倾斜并向上延伸到露台的透明幕墙。在进入建筑后,可扩展空间带来的强烈冲击感在三层高的中庭内立即显现出来。一条竖向阶梯从建筑的中央一直升至顶部,直达戏剧性的屋顶露台。

Azabu-Jyuban

Tokyo, Japan
1987
K-One Corporation

In this project, Zaha Hadid Architects proposed a means to liberate space in Tokyo, where most sites are tightly constrained due to limited available space, and many buildings only contribute to the city's densely populated environment.

Carving through the landscape and burrowing into the depths, the Azabu-Jyuban building purposefully amplifies the pressure from its narrow site, which is situated in a canyon of diverse structures in the vicinity of the Roppongi district. The pristine glass structure of the Azabu-Jyuban building is compressed between a towering metal wall and a reinforced concrete wall, both adorned with jewel-like windows. Within this space, two curtain walls of Azabu-Jyuban emerge—a striking blue glass one and a transparent one—which lean outward and ascend to the parapet walls of the terrace. Upon entering the building, the profound impact of the expandable space becomes immediately evident within the three-storey atrium. A vertical staircase ascends from the building's heart all the way to the top, exploding into dramatic balconies.

季风餐厅

日本，札幌
1989—1990年
Jasmac Corporation

冰与火的交融，在季风餐厅内催生出两处迷人的场所。针对正式用餐区和休闲酒廊，ZHA精心设计了对比鲜明的两种空间氛围。

从札幌传统的冰晶结构中汲取灵感，一楼的正式用餐区采用宁静的冷灰色配色，并以玻璃和金属材料为主。桌子被重新定义为光滑的冰块，升高的地板则像冰山一样蜿蜒穿过整个空间。

在用餐区之上，休闲酒廊充斥着鲜亮的红色、灿烂的黄色和艳丽的橙色。吧台上方的设计呈螺旋状上升至天花板，就像是一股热旋风冲破了受压的容器。为了提升与完善餐厅体验，ZHA设计了一系列仿生形态沙发。它们有着多种座位类型，并配有可移动餐盘及与之适应的"嵌入式"沙发靠背。

Moonsoon Restaurant

Sapporo, Japan
1989–1990
Jasmac Corporation

Fire and ice converged to give birth to two captivating realms within the Moonsoon Restaurant. Tasked with designing an interior that encompasses both a formal dining area and a relaxation lounge, Zaha Hadid Architects crafted a compelling interplay of contrasting atmospheres.

Drawing inspiration from the traditional ice structures of Sapporo, the ground floor of the formal dining area featured a serene cool grey colour scheme, embodied by glass and metal materials. Tables were reimagined as sleek fragments of ice, while a raised floor level meandered like an iceberg across the space.

Above the icy chamber, the relaxation lounge blazed with an inferno of vibrant reds, dazzling yellows and exuberant oranges. A spiralling feature above the bar pierced through the ceiling of the ground floor, curling up to the underside of the domed upper level like a fiery whirlwind breaking through a vessel under pressure. To enhance and complete this restaurant experience, a plasma-like collection of biomorphic sofas allowed for various flexible seating types, accompanied by movable trays and adaptable 'plug-in' sofa backs.

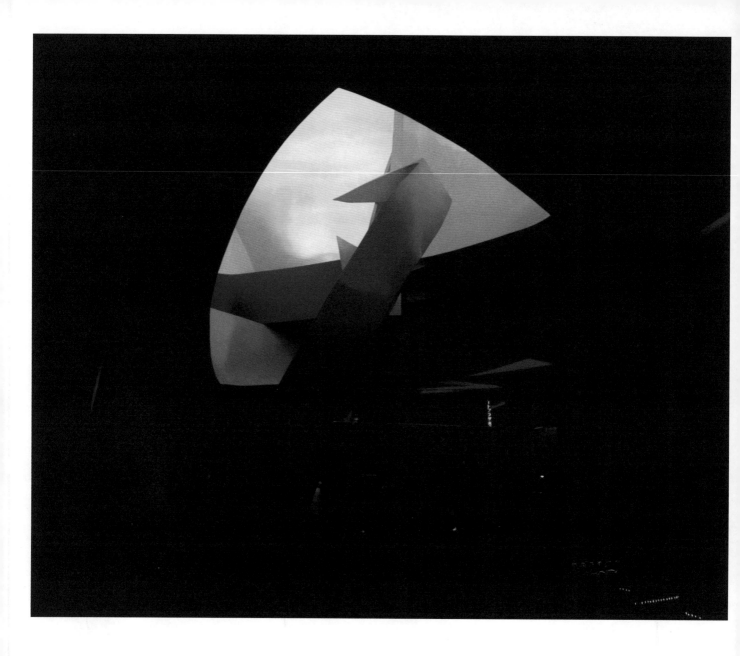

维特拉消防站

德国，莱茵河畔威尔城
1991—1993年
Vitra International AG

经过对莱茵河畔威尔的维特拉工厂整体场地的研究，ZHA将维特拉消防站设计为线性景观区内的关键元素。维特拉消防站作为邻近田地与葡萄园中线性模式的建筑性延伸，是连接体，而非孤立个体。

将功能延伸到一个狭长的建筑中，界定并屏蔽工厂场地的边界，这成为建筑概念的出发点：一个线性的、分层的系列墙面。建筑栖息在墙体间的空间中，这些墙体根据功能要求"穿孔""倾斜"与"隔断"。从正面看，该建筑呈密封状态，其内部空间仅可从高空窥见。人们经过时会瞥见红色的消防车和消防员们进行日常锻炼。建筑中"凝固的运动"，表达了消防站始终处于警戒状态的紧张氛围。

整体建筑采用裸露的钢筋现浇混凝土建造。设计特别强调所有边缘的锋利度。为强调建筑概念的抽象性及棱柱形式的简洁性，设计摒弃了如屋顶边缘及覆层等所有附加元素。这种消隐的细节处理被转译为无框玻璃、包围车库的大型滑动平面及内部空间照明方案。光线可引导使用者在建筑中精准地移动。

Vitra Fire Station

Weil am Rhein, Germany
1991–1993
Vitra International AG

The design for the Vitra Fire Station was initiated with a study of the overall Vitra factory site in Weil am Rhein. The building is located within a linear landscaped zone, which is an artificial extension of the linear patterns in adjacent fields and vineyards. Conceived as the key element, the building was designed as a connecting unit rather than an isolated object.

Stretching the programme into a long, narrow building defines and screens the border of the factory site, which became the point of departure for the architectural concept: a linear, layered series of walls. The programme of the architecture inhabits the spaces between these walls, which 'puncture', 'tilt' and 'break' according to functional requirements. But the building is hermetic from a frontal reading, revealing the interiors only from a perpendicular viewpoint. As one passes across the spaces, one catches glimpses of the red fire engines and exercising rituals of the firemen. These frozen movements in the architecture resemble a series of choreographic notations, expressing the tension of always being on the alert.

The whole building is constructed of exposed reinforced in-situ concrete. Special attention was given to the sharpness of all edges; any attachments like roof edgings or claddings were avoided as they distract from the simplicity of the prismatic form and the abstract quality of the architectural concept. This same absence of detail informed the frameless glazing, the large sliding planes enclosing the garage, and the treatment of the interior spaces including the lighting scheme. The lines of light direct the necessarily precise and fast movement through the building.

费诺科学中心

德国，沃尔夫斯堡
2000—2005年
Neulandgesellschaft mbH
On behalf of the City of Wolfsburg

费诺科学中心位于沃尔夫斯堡市的重要位置，此处既是多个重要文化建筑（由阿尔托、沙隆和施维格设计）的聚集地，也是连接米特兰卡纳尔（大众汽车城）北岸的纽带。该中心的建筑理念来源于魔盒的概念——一个一打开就能唤醒人们的好奇心和探索欲望的盒子。

多条人行道与车道延伸至该处，在人工景观与建筑内部有效构建出运行路径的界面。从体量上看，建筑结构最大限度地保持了建筑的透明性与多孔性。主要的展览空间被抬升，以覆盖室外公共广场，广场以混凝土锥体结构为支撑，提供各种商业和文化功能。

展览空间内打造了一个类似火山口的空间形态，使不同高度的展览空间之间可以产生斜向的视线连接。访客会发现自己置身于一个由250个"实验站"组成的奇境中，这些"实验站"突出的体量可适应各种后台功能的需要。

Phaeno Science Centre

Wolfsburg, Germany
2000–2005
Neulandgesellschaft mbH
On behalf of the City of Wolfsburg

The Phaeno Science Centre occupies an important site at the centre of Wolfsburg. It completes a chain of significant cultural buildings by Aalto, Scharoun and Schweger, and connects via a bridge across the Mittelland Kanal to Volkswagen's car-manufacturing city, Autostadt. The architectural concept of the building was inspired by the idea of a magic box—an object that awakens curiosity and the desire for discovery upon opening.

Multiple threads of pedestrian and vehicular movement are pulled through an artificial landscape and into the building. Volumetrically, the building is structured to maximise the sense of transparency and porosity. To achieve this, the main volume that houses the exhibition space is raised to cover an outdoor public plaza that accommodates a variety of commercial and cultural functions, housed within the supporting concrete cones.

An artificial crater-like landscape is developed inside the exhibition space, allowing diagonal views onto different exhibition levels. Visitors find themselves in a wonderland of 250 Experimental Stations in protruding masses that accommodate a variety of back-of-house functions.

香奈儿当代移动艺术馆

中国，香港；日本，东京；美国，纽约
2008年
香奈儿

这个700平方米的当代移动艺术馆是为香奈儿打造的一个独特的雕塑式展览和活动空间。展馆采用钢和铝混合结构，配有PVC外表皮和ETFE软膜，可在不同地点高效拆卸和重新安装。表面精致、流畅的层次细节构成了一个优雅连贯的整体。一系列连续的拱门围绕着中央庭院，形成了一个多功能建筑结构。

灯光透过透明天花将墙面照亮，突出"拱形"结构，并为艺术馆营造出一种新颖的空间体验。宽敞的屋顶天井让入口沐浴在自然光线中，从而模糊室内外空间的界限。65平方米的中央庭院向天空敞开，形成展览和公共区域的过渡空间。反光材料可使外表皮呈现出不同的色彩，以适应各种活动的需要。

Mobile Art — Chanel Contemporary Art Container

Hong Kong, China; Tokyo, Japan;
New York, USA
2008
Chanel

The 700-square metre Mobile Art Container is a unique sculptural pavilion created as an exhibition and event space for Chanel. The container's structure is designed in steel and aluminium with PVC cladding and ETFE lights, allowing for efficient disassembly and reinstallation in different locations. Its smooth layering of exquisite details creates an elegant and cohesive whole, resulting in a versatile architectural structure of a series of continuous arches sequencing towards a central courtyard.

Artificial light behind the translucent ceiling washes the walls to emphasise the 'arched' structure and assists in the creation of a new artificial landscape for art installations. The expansive roof light opening bathes the entrance in natural daylight, blurring the boundaries between interior and exterior spaces. The 65-square metre central courtyard, largely open to the sky above, serves as an intermediate space between exhibition and public areas. Reflective materials allow the exterior skin to be illuminated with varying colours, which can be tailored to the various programmes of special events.

MAXXI：
国立二十一世纪美术馆

意大利，罗马
1998—2009年
意大利文化部

"MAXXI：国立二十一世纪美术馆"的设计以罗马的城市面貌为灵感，呈现出一种线性语言。建筑沿相邻的城市肌理向三个方向延伸，使馆内流线和布局充分响应周边城市。

ZHA并没有从概念上将美术馆定位为一个孤立的主体，而是将展厅和流线设计成交织在一起的空间元素。美术馆前方的空间被当地居民看作新的罗马广场，从而成了罗马城市规划中不可或缺的一部分。美术馆的设计与城市的紧密关系在这里得以彰显。建筑内部的流线顺应周围的城市流线而设，各条流线和谐共舞，直到分离、交错，或转折到L形场地的角落。它们会集的流线元素是垂直的、倾斜的，它们构建的空间是内外交织的，它们组合的展厅是紧密而广阔的。美术馆内部的五个展厅被巧妙地布置成三个环形，游客可以沿着不重复的路径再次回到大厅。

美术馆的多孔性与沉浸式设计使室内空间充满了艺术氛围，同时表明了建筑与其所承载的活动之间的交融。ZHA的设计直面自20世纪60年代以来艺术实践引发的物质和概念上的冲突，摒弃、割舍了以"物体"为主要导向的画廊空间，转而强调艺术、建筑与人之间的动态关系。与传统设计方法不同的是，这种设计不再把个别艺术品视为神圣、不可亵渎的，而是寻求对艺术的某种全新理解——将其视为更加宏观的动态领域的内在组成部分。

MAXXI: Museum of XXI Century Arts

Rome, Italy
1998-2009
Italian Ministry of Culture

The MAXXI Museum was designed and developed as a linear system drawn from the surrounding urban fabric of Rome. Extending the adjacent urban grids onto the site from three directions, the urban fabric directly informed the interior circulation/organisation of the museum.

Rather than situating the museum as a singular object, the museum galleries and circulation were conceived as a field of interwoven spatial elements. The extent of this connection to the city is demonstrated in front of the building where the local community uses the space as a new Roman piazza, an integral part of the urbanism of Rome itself. The circulation within the museum follows the surrounding urban movement, with the lines flowing in unison until they separate, intersect or turn the corner of the L-shaped site. They form circulation elements—vertical and oblique, spaces—interior and exterior, and galleries—intimate and expansive. The interior of the building features five galleries that are arranged in three loops, allowing visitors to return to the lobby without retracing their steps.

The porous and immersive nature of the museum is indicative of the relationship between the architecture and the activities it will host. The design confronts the material and conceptual dissonance evoked by art practice since the late 1960s. It promotes a disinheriting of the 'object'-oriented gallery space and instead highlights the dynamic agency between art, architecture and people. Departing from the traditional approach of treating individual art objects as sacredly untouchable, the design seeks a new understanding of art as an intrinsic piece within a larger dynamic field.

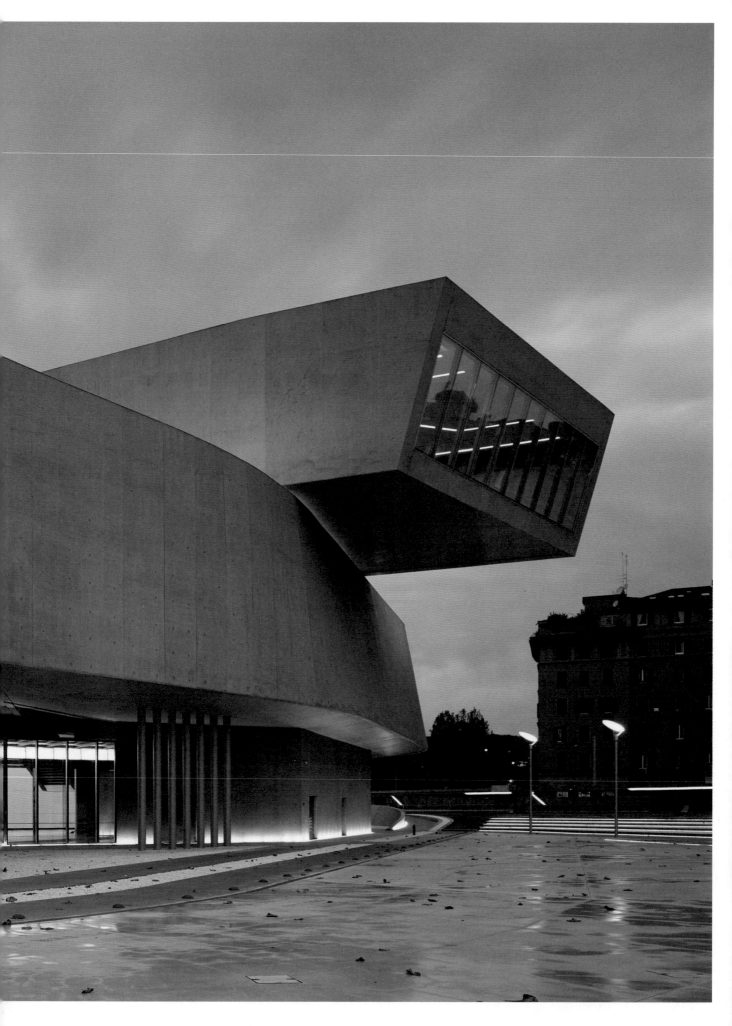

伦敦水上运动中心

英国,伦敦
2005—2011年(奥运模式)
2014年(遗产模式)
伦敦奥运交付管理局

伦敦水上运动中心的设计既满足了2012年伦敦奥运会的规模和容量要求,又为奥运会后的遗产模式提供了最佳的匹配方案。当该中心2014年向公众开放时,英国《卫报》这样报道:"……世界上最令人惊叹的城市泳池。从用草包围低层的弧形墙到波浪形的屋顶,这个有机的结构在向你招手……直到你到达大教堂般的泳池大厅内部,你才意识到建筑师的卓越技巧。如今,任何人都可以在这里游泳,其门票价格与本地其他游泳池的价格相近。这是奥运会给我们带来的一项伟大的遗产。"

在2012年奥运会与残奥会之后,水上运动中心过渡到遗产模式。玻璃幕墙取代了原有的临时座位,减小了泳池大厅的体积,但保持了2800个座位的容量供社区和国家或国际活动使用。2014年以来,超过800万不同年龄段、不同运动能力的人,来到水上运动中心游泳。它已经成为伦敦东部社区一个受人喜爱的公共空间与运动场所。

London Aquatics Centre

London, UK
2005–2011(Olympic mode)
2014(Legacy mode)
Olympic Delivery Authority

The London Aquatics Centre was designed to both accommodate the size and capacity of the London 2012 Olympic Games while also providing the optimum size and capacity for use in Legacy mode after the Games. On opening to the public in 2014, the centre was reported by the Guardian newspaper as '…the most jaw-dropping municipal swimming pool in the world. From the curved wall of grass that encloses the lower levels to the swooping wave of a roof, this organic-looking structure beckons you in… it is not until you reach the cathedral-like interior of the main pool hall that you realise the extraordinary skill of the architect. Anyone can now swim in it for around the same entrance fee as other local pools. It is a great legacy of the Olympics.'

Following the 2012 Olympic and Paralympic Games, the Aquatics Centre transitioned into its Legacy mode. Temporary seating was replaced with glazing panels, reducing the pool hall volume but maintaining a capacity of 2,800 seats for community use and future national/international events. Since 2014, over 8 million people, spanning various age groups and abilities, have visited the Aquatics Centre for swimming. It has become a beloved public space and a popular sporting venue for the community in East London.

阿利耶夫中心

阿塞拜疆，巴库
2007—2012年
阿塞拜疆共和国政府

阿利耶夫中心已经成为阿塞拜疆重要的文化建筑。其设计灵感来自阿塞拜疆丰富的伊斯兰历史建筑遗产，其中建筑元素与周围环境之间的流动性是一个被反复强调的主题。设计旨在通过细致入微的理解对阿塞拜疆历史文化进行当代诠释。

该中心的设计使周围广场与建筑内外空间建立起了一种连续、流动的关系。广场作为巴库城市肌理的一部分，向城市开放。室外广场的地面逐渐升起，围合出建筑空间并在室内延伸打造了一系列公共活动空间。这些空间共同展示着阿塞拜疆的当代与传统文化。设计通过起伏、分岔、褶皱与转折等多变的空间形式将阿利耶夫中心进一步阐述为一种建筑景观，欢迎、接纳和引导游客在不同层次的内部空间中体验。通过这种姿态，该建筑模糊了建筑主体与城市景观、建筑外幕墙与城市广场、室内与室外之间的传统界限。

Heydar Aliyev Centre

Baku, Azerbaijan
2007–2012
The Republic of Azerbaijan

The Heydar Aliyev Centre was designed to be the primary building for cultural programs in Azerbaijan. Its design draws inspiration from the rich heritage of Azerbaijan's historical Islamic architecture, in which the fluidity between architectural elements and their surrounding environment is a recurring theme. Relating to this historical understanding of architecture, the intention of the design was to develop a contemporary interpretation of Azeri culture that would reflect a more nuanced understanding.

The design of the Centre establishes a continuous, fluid relationship between its surrounding plaza and the building's interior. The plaza serves as the ground surface and is accessible to all as an integral part of Baku's urban fabric. It rises to envelop an equally public interior space and define a sequence of event spaces dedicated to the collective celebration of contemporary and traditional Azeri culture. The plaza is further elaborated into an architectural landscape through formations such as undulations, bifurcations, folds, and inflections. These transformations enable it to perform various functions of welcoming, embracing, and directing visitors through different levels of the interior. With this gesture, the building blurs the conventional differentiation between architectural object and urban landscape, building envelope and urban plaza, interior and exterior.

东大门设计广场

韩国,首尔
2007—2014年
首尔市政府,首尔设计基金会

东大门设计广场被打造为一处建筑景观,其设计以古城墙与新出土的文物为灵感,与东大门历史文化公园紧密相连,共同打造出一个独特的空间。它处于首尔最繁忙、历史最悠久的一个区域。作为该区域的文化中心,东大门设计广场为各个年龄段的公众提供了一个聚集的场所。作为促进思想交流,以及新技术与媒体探索的催化剂,广场持续不断地举办展览与活动,为城市注入文化活力。

广场设计中的流动形态将广场与城市巧妙地融合为一体,鼓励首尔市民与广场之间的互动。该设计模糊了建筑物与自然之间的界限,并且匠心独具地将城市、广场与建筑连接在一起。通过建筑空间的透空和转折,游客得以一窥独具一格的设计世界。该设计使得东大门设计广场成为首尔当代文化、高速发展与历史之间重要的纽带。

Dongdaemun Design Plaza

Seoul, Korea
2007-2014
Seoul Metropolitan Government,
Seoul Design Foundation

The Dongdaemun Design Plaza (DDP) is an architectural landscape that revolves around the ancient city wall and newly discovered cultural artefacts, which form the central element of the composition linking the park and plaza together. Designed as a cultural hub at the heart of one of the busiest and most historic districts of Seoul, the DDP is a place for people of all ages. Serving as a catalyst to instigate the exchange of ideas and exploration of new technologies and media, the DDP hosts ongoing exhibitions and events that bring cultural vitality into the city.

The fluid shape of the design integrates the plaza and the city seamlessly as one and encourages interactions between the plaza and the people of Seoul. It blurs the boundary between architecture and nature and connects the city, plaza, and architecture in an innovative way. Voids and inflections in the plaza's surface give visitors glimpses into the innovative world of design below, making the DDP an important link between the city's contemporary culture, emerging nature, and history.

丽敦豪邸

新加坡
2007—2014年
凯德集团

丽敦豪邸是一个高层住宅项目，位于新加坡市第十区中心，与其他商业区和市政中心有便利的交通联系，并根据居民的特定需求提供各式各样的公寓布局。该住宅项目共有1715个住宅单元，分布在7座塔楼和12栋半独立式别墅中。该设计巧妙地将场地中的挑战转化为优势，并基于每套公寓在场地中的位置和条件进行量身定制。

塔楼由上至下向内收窄，形成独特的花瓣形布局。这个由ZHA设计的花瓣形布局优化了底层公共空间的利用，并为每套公寓提供三面采光，确保所有厨房和浴室的自然通风。建筑的朝向和位置考虑了当地环境强烈的日照环境的影响，并最大限度地扩展了面向城市景观的视野。该设计将周围的城市轴线延伸至场地内，生成一系列线条来定义不同的景观主题，在场地内创造多样化的体验。

根据每层住宅单元的数量，每座塔楼被细分为4~8个花瓣，从而提供多元的公寓产品。花瓣形的平面布局随着塔楼的高度变化而逐渐改变形状，以适应不同配置和类型的住宅单元。这样的设计使塔楼能够适应场地条件、内部组织和结构优化所产生的一系列参数变化。

D'Leedon, Singapore

Singapore, Singapore
2007–2014
CapitaLand-led consortium

Located in the centre of Singapore's District 10 and well-connected to other commercial and civic centres, D'Leedon is a high-rise residential development that offers highly differentiated apartment layouts according to the residents' specific needs. The residential programme accommodates 1,715 units, distributed in 7 towers and 12 semi-detached villas. The design turns the challenges of the site into advantages for residents and customises the design of single apartments based on their locations and conditions within the site.

The towers taper inwards as they reach the ground and form a unique petal-shaped layout. Developed by Zaha Hadid Architects, this petal-shaped form optimises the ground-level public space utilisation, and provides windows on three sides of every apartment, along with natural ventilation in all kitchens and bathrooms. The orientation and placement of the buildings are designed in response to environmental considerations of managing intense sunlight and to maximise views across Singapore. Extending the surrounding alignments of axis into the site, a series of lines are generated to define different landscape themes, creating diverse experiences throughout the whole site.

According to the number of residential units per floor, each tower is subdivided into four to eight petals, resulting in a large diversity of apartments. The generative floor plan of the petal changes shape along the height of the tower to accommodate different configurations and types of residential units. The changing composition of unit type allows the towers to respond to a series of parameters dictated by site conditions, internal organisation, and structural optimisation.

阿卜杜拉
国王石油研究中心

沙特阿拉伯,利雅
2009—2017年
沙特阿拉伯国家石油公司

阿卜杜拉国王石油研究中心(KAPSARC)是一个非营利性质研究机构,致力于能源高效利用的开发。这座占地70,000平方米的园区包括五座建筑:能源知识中心、能源计算中心、会议中心(内设展厅和礼堂)、科研图书馆和穆斯林祷告堂。该设计以扎实的技术和环境考量为基础,将园区的五种元素融合为一个统一的整体。

作为ZHA首个获得美国绿色建筑委员会LEED白金认证的项目,该园区在设计上充分考虑了如何应对利雅得的高原环境条件,以最大限度地减少能源和资源消耗。该设计的主要策略是打造一个蜂窝状结构——由单个单元组成的模块化系统,从而将不同的建筑体块融合为一个整体,并提供相互连接的公共空间。蜂窝状结构最大限度地优化了功能、空间和结构的效率和一致性,同时减少了材料使用。蜂窝状结构设计还便于根据未来需求进行灵活调整,以及通过加建额外单元进行扩展和整合。

该园区的设计营造了高度的开放感和参与感,旨在鼓励研发人员和访客之间的积极交流。通过巧妙的错层布局,室内设计创造了通透的空间层次,特别是用作非正式会议的协作空间,促进研发人员相互交流;私密空间则位于楼层重叠的地方。

King Abdullah Petroleum Studies and Research Centre

Riyadh, Saudi Arabia
2009–2017
Saudi Aramco

King Abdullah Petroleum Studies and Research Centre (KAPSARC) is a non-profit institution dedicated to independent research on effective energy-use policies. The 70,000 square-metre KAPSARC campus incorporates five buildings: the Energy Knowledge Centre, Energy Computer Centre, Conference Centre with exhibition hall and auditorium, Research Library and the Musalla, a place for prayer. With solid technical and environmental considerations at heart, the design draws the five elements of the campus into a unified whole.

As ZHA's first project to be awarded LEED Platinum certification by the US Green Building Council, the campus is designed in response to the environmental conditions of the Riyadh Plateau to minimise energy and resource consumption. The primary organising strategy of the design is a partially modular system consisting of individual cells. This system allows for different departmental buildings to integrate as a single ensemble with interconnecting public spaces. It maximises organisational, spatial and structural efficiency and consistency, and it minimises material usage with hexagonal honeycomb structures. The honeycomb grid design also allows for easy adaptations for future requirements and incorporation of additional cells for future expansion of the campus.

The design fosters a sense of openness and engagement, facilitating an active exchange between researchers and visitors. By strategically off-setting floorplates, the interior creates spatial layering effects that allow transparency between floors, particularly in areas designed as collaborative zones for informal meetings. Areas requiring privacy are located at where floorplates overlap.

忠利集团大厦

意大利，米兰
2004—2018年
City Life Consortium

忠利集团大厦是米兰CityLife总体规划的一部分，这是在2005年新米兰展览中心迁移至Rho Pero之后重新开发的一处商业广场。忠利集团大厦最多可容纳3900名员工，为全球最大的金融机构之一的持续发展提供支持。这座170米高（44层）的大厦与周围的公共广场和公园相连接，并呼应该城市的三条主要轴线。

  塔楼裙房的曲线几何造型是由围绕向心力旋涡的扭曲而定义的。这种向心力来自塔楼底部三条城市轴线的交错相会。地面上的这种向心力旋涡通过重新排列的菱形楼层垂直穿过整栋建筑，并使大楼围绕着垂直轴旋转。这种螺旋形的扭曲随着楼层的增高而逐渐减弱，同时也让上下楼层之间的关系略有不同。随着楼层升高，视野逐渐开阔，塔楼的高层与城市的东南轴线对齐，朝向15世纪布拉曼特重修的圣玛丽亚修道院。

  该项目不仅严格遵守当地的建筑法规，而且也是能源高效利用的国际典范。带有遮阳百叶窗的双层玻璃幕墙为每层楼提供高效的环境控制，确保了卓越的节能效果，也使大厦获得了美国绿色建筑委员会的LEED白金认证。

Generali Tower

Milan, Italy
2004–2018
City Life Consortium

Generali Tower is part of the CityLife masterplan, which redeveloped Milan's abandoned trade fair grounds following the fair's relocation to Rho Pero in 2005. The tower accommodates up to 3,900 employees to support the continuous growth of one of the world's largest financial institutions. Aligned with three primary axes of the city that converge within CityLife at ground level, the 170-metre (44-storey) Generali Tower connects seamlessly with the surrounding public piazzas and park.

  The curvilinear geometries of the tower's podium are defined by a twist around a vortex of centripetal forces. These forces result from the staggered intersection of the three city axes at the tower's base. The vortex of centripetal forces at ground level is transferred vertically through the tower by realigning successive rhomboid-shaped floor plates, twisting the tower about its vertical axis. This helical twist gradually diminishes with each floor, giving all floors a fractionally different relationship to the floors above and below. As the tower rises, offering expansive views of Milan, the tower aligns its higher floors with the city's southeast axis, leading to Bramante's 15th Century tribune of Santa Maria della Grazie.

  The tower excels in all international benchmarks for efficiency while respecting Milan's rigorous local building codes. Its double-façade, consisting of sun-deflecting louvers flanked by glazing, provides exceptional environmental control for each floor, ensuring excellent energy performance. These features contribute to the Generali Tower's LEED Platinum certification.

纽约公寓

美国，纽约
2013—2018年
Related Companies

纽约的城市经过一代代发展变迁，在街道和高线公园之间形成了层次丰富的城市景观，造就了强大的城市活力。ZHA将历史文化与创新的设计理念相结合，融入纽约公寓的设计，对该地区的深厚历史做出了全新演绎。

该公寓的设计回应了周围城市空间的多层次特性，用错落有致的楼层定义了生活空间。手工打造的金属幕墙形成交错的V形图案，突出别致的楼层分割方式。其精湛的工艺也反映出切尔西的工业历史，体现了纽约历史建筑公共空间的悠久传统。建筑外墙的设计和建造都基于对材料及技术的成熟运用。手工着色的外墙突显了纽约公寓对细节的专注，与高线公园及周围建筑群形成了呼应。

这座11层高的公寓可容纳39个住宅单元，每个单元都配有由扎哈·哈迪德设计（ZHD）为其量身打造的波菲厨房，同时还设有自动化代客泊车和储藏系统。建筑有多个核心筒，以提供尽可能多的私人电梯。公寓内部设施包括带有水疗的健康中心、23米长带天窗的游泳池、雕塑花园以及配有IMAX影院的娱乐空间。纽约公寓所在的社区拥有350多间艺术画廊，见证了高线公园从废弃货运铁路向公共空间的巨大转变。纽约公寓的设计维护了其所在的社区的鲜明特性，使其在保有自身建筑风格的同时，与周围环境和谐相融。

520 West 28th

New York City, USA
2013–2018
Related Companies

The powerful urban dynamic between the streets of New York and the High Line results in a layered civic landscape that has developed over generations. The architecture of 520 West 28th embraces this contextual relationship and employs new ideas and concepts to propel the site's rich history into its latest evolution.

Echoing the layered nature of the surrounding civic spaces, the design of 520 West 28th defines living spaces through split levels. These split levels are emphasised by the interlocking chevron pattern on the hand-crafted steel façade. Its meticulous craftsmanship reflects the industrial history of Chelsea and the venerable tradition of enhancing public spaces in New York's historic architecture. Designed and constructed with a practiced understanding of material qualities and manufacturing techniques, the façade is brushed and tinted by hand and conveys the attention to detail throughout 520 West 28th and its neighbourhood.

The 11-storey building houses 39 residences, each featuring tailored interiors with Boffi kitchens by Zaha Hadid Design and integrated technologies including automated valet parking and storage. Designed with multiple cores, the building offers most residences private elevator lobbies and amenities, such as a wellness level with spa, a 23-metre sky-lit lap pool, a sculpture garden and an entertainment suite with an IMAX theatre. Located in an established community with over 350 galleries, the building witnessed the neighbourhood's evolution from an abandoned freight rail line to a public park. Upholding the distinctive character of its neighbourhood, 520 West 28th represents a building of its own architectural identity yet very much in harmony with its surroundings as well.

贾努布体育场

卡塔尔，沃克拉
2014—2019 年
卡塔尔交付和遗产最高委员会

作为2022年卡塔尔世界杯的第一个新体育场，贾努布体育场采用由德国施莱希工程设计公司设计的可伸缩屋顶，为比赛场地提供遮蔽。ZHA与Aecom合作，于2013年3月开始设计该体育场及周围的新城区。体育场内还整合了一套太阳能冷却系统，以确保场馆在卡塔尔炎热的夏季仍可使用。该体育场采用被动式设计原则，并经过详细的计算机建模和风洞测试，旨在最大限度地提升围护结构的有效性，从而为球员与观众提供舒适的环境。

场馆的设计灵感来源于当地传统的单桅三角帆船，以此致敬沃克拉市的航海历史。抽象的设计回应了气候、环境和足球场功能需求的现实条件。

2022年世界杯结束后，该体育场的40,000个座位将减少到20,000个，这是当地职业球队沃克拉足球俱乐部主场的最佳容量。临时座位采用了可拆卸的座椅设计，以便赛后运输到需要体育基础设施的国家和地区。

Al Janoub Stadium

Al Wakrah, Qatar
2014–2019
Supreme Committee for
Delivery and Legacy of
the 2022 FIFA World Cup Qatar ™

As the first new stadium commissioned for the 2022 FIFA World Cup Qatar, the Al Janoub stadium incorporates an operable roof, designed by Schlaich Bergermann Partner, to shade the field of play. Working with Aecom, Zaha Hadid Architects (ZHA) began designing the stadium along with its new precinct for the city in March 2013. A solar-powered cooling system is also integrated into the stadium to ensure its functionality during Qatar's hot summer months. Passive design principles, along with detailed computer modelling and wind tunnel testing, were employed to maximise the effectiveness of the enclosure for the comfort of players and spectators.

Reflecting Al Wakrah's maritime heritage, the stadium's design incorporates local cultural references to the traditional boat of the region, the dhow. The abstracted design is combined with practical responses to climate, context and functional requirements of a football stadium.

The stadium's 40,000-seat capacity for the 2022 World Cup will be reduced to 20,000 seats in its legacy mode following the tournament, which is the optimum capacity as home ground to the local Al Wakrah Sport Club professional team. The temporary seats have been designed to be demountable for transportation to a country in need of sporting infrastructure.

奥普斯大厦

阿联酋，迪拜
2012—2020年
迪拜奥美地产

奥普斯大厦位于迪拜哈利法塔所在区。该建筑为两座塔楼连接而成的一个立方体。立方体的中心经过"雕琢"后，形成一个流线型的曲面孔洞，这也是建筑空间中重要的一部分。站在中央位置可以将建筑的外部景观尽收眼底。八层高的曲面孔洞充满流动感，与周围精确的正交几何体形成了鲜明的对比。

这两座塔楼在地面由一个四层高的中庭相连，在高层则由一座不对称的三层高的连桥连接。这座非对称的连桥距地面71米，横跨三层楼，宽达38米。奥普斯大厦内部功能丰富，里面有闻名遐迩的迪拜梅里亚酒店，还有各种餐厅、办公空间与住宅公寓。

Opus

Dubai, UAE
2012–2020
Omniyat Properties LLC

Located within Burj Khalifa district of Dubai, the Opus is designed as two separate towers that coalesce into a singular whole—taking the form of a cube. The cube is then 'carved', creating a central void that is an important volume within the building in its own right—providing views to the exterior from the centre of the building. The free-formed fluidity of this eight-storey void contrasts with the precise orthogonal geometry of the surrounding cube.

The two towers are linked by a four-storey atrium at ground level and are also linked by a connecting bridge at the upper floors 71 metres above the ground. This asymmetric bridge spans three storeys and measures 38 metres in width. Within the Opus, one can find the distinguished ME Dubai hotel, a variety of restaurants, office spaces, and residential apartments.

千号馆

美国,迈阿密
2012—2020年
1000 Biscayne Tower, LLC

千号馆坐落于迈阿密比斯坎大道上,是一座62层的住宅公寓。其标志性的外骨骼结构设计使它拥有十分宽敞、几乎无柱的内部空间。这种外骨骼结构的设计和技术创新延续了ZHA对高层建筑的研究,使流畅的建筑形式表达与先进的工程学形成高度统一的关系。

与传统塔楼构造不同,千号馆的外骨骼从裙房上方开始"生长",随着塔楼纤长的体量蜿蜒上升,外部的玻璃立面保持了高度的连贯性和统一性。随着阳台层层叠叠的节奏变化,玻璃护栏将建筑围合成一个闪亮的多面晶体,与亚光的混凝土结构框架形成鲜明对比,也使千号馆在当地不断变化的光照条件下充满活力。

这个高216米的混凝土外骨骼结构可以抵御迈阿密季节性飓风所产生的高达290千米每小时的风速。对于实现如此的高度、精度及要求苛刻的建造计划,协作、测试和创新的技术解决方案至关重要。为了应对这一挑战,世界上第一个使用玻璃纤维增强混凝土(GFRC)永久模具的建筑系统被开发了出来。工厂预制的GFRC板,作为混凝土浇筑模具的同时也是建筑的外表皮,在逐层围绕钢筋组装后,被注入高强混凝土。

One Thousand Museum

Miami, USA
2012–2020
1000 Biscayne Tower, LLC

One Thousand Museum stands prominently as a 62-storey residential tower along Biscayne Boulevard in Miami. With its significant structural frame at the perimeter, the interior allows for living spaces uninterrupted by internal columns. This design and technical innovation of the exoskeleton structure represents a line of Zaha Hadid Architects' research in high-rise construction, integrating fluid architectural expression with advanced engineering.

Unlike a conventional configuration of a tower resting on a base, the exoskeleton frame emerges from the podium, rising sinuously over and around the slender volume of the tower and envelops its glass façade through one continuous, unifying feature. With varying balcony recesses, the reflective glass enclosure resembles a multi-faceted crystal, starkly contrasting the matte and cementitious structural frame, a contrast that animates the tower under Miami's continuously changing light conditions.

The 216-metre tall concrete exoskeleton structure is designed to withstand winds of up to 290 kilometres per hour due to Miami's seasonal hurricanes. Collaboration, testing, and an innovative technical solution were crucial to achieving construction of such precision and height, as well as demanding construction program. To face this challenge, a world-first construction system was developed using Glass Fibre Reinforced Concrete (GFRC) permanent formwork. Factory-made GFRC panels, which provide the formwork and the architectural finish, are assembled level by level around steel reinforcement cages, and then filled with high-strength concrete.

碧哈总部

阿联酋，沙迦
2014—2022年
碧哈集团

碧哈集团业务覆盖六个重点行业，包括废弃物管理、废弃物回收、清洁能源、环境咨询、教育和绿色出行。碧哈总部项目是该集团的最新里程碑，它将持续引领沙迦乃至全世界的创新工程。新总部由太阳能电池板提供电力，按照LEED白金标准运行，旨在实现净零排放和最低的能源消耗，为未来的办公建筑树立新的标杆。

总部的设计体现了碧哈的可持续发展与数字化原则，通过一系列相互连接的"沙丘"回应场地环境。这些"沙丘"的朝向和形状都是为了优化当地气候条件。该设计植根于沙迦的阿萨贾沙漠，与四周被盛行风塑造而成的沙丘和山脊相呼应。

在员工体验方面，该项目设置了免接触式通道、虚拟接待、智能会议室，以及可以将日常工作自动化的辅助应用程序。建筑的智能管理系统能够根据一天中的时间变化和使用人数调节光照和温度。办公室还可适应远程办公、混合办公等不同工作场景。

BEEAH Headquarters

Sharjah, UAE
2014–2022
BEEAH Group

Working across six key industries, the endeavours of BEEAH Group include waste management, waste recycling, clean energy, environmental consulting, education and green mobility. The BEEAH headquarters is the latest milestone for the Group as it continues to pioneer innovations for Sharjah and across the globe. Powered by its solar array for operations at LEED Platinum standards, the new Headquarters has been designed to achieve net-zero emissions with minimal energy consumption, setting a new benchmark for future workplaces.

Embodying BEEAH's key principles of sustainability and digitalisation, the design of the headquarters responds to its environment as a series of interconnecting 'dunes,' orientated and shaped to optimise local climatic conditions. Embedded within its context of Sharjah's Al Sajaa desert, the design echoes the surrounding landscape, shaped by prevailing winds into concave sand dunes and intersecting convex ridges.

The employee experiences include contactless pathways, a virtual concierge, smart meeting rooms and a companion app that automates day-to-day tasks. The smart management system of the building automatically adjusts lighting and temperature depending on occupancy and time of day. The rooms are also equipped for remote and hybrid work scenarios.

水星大厦

马耳他，佩斯维尔
2016—2023年
J. Portelli Projects

水星大厦的改造和重建工作将住宅公寓和精品酒店融入了马耳他充满活力的城市环境之中。佩斯维尔的悠久历史可以追溯到20世纪初，该地区自20世纪60年代以来一直是闻名遐迩的旅游胜地，拥有众多餐馆、酒吧、夜总会、赌场、码头和国际酒店。该项目为岛上居民和游客提供了全新的城市空间和便利设施。通过改善有限的住房供应条件，加大对公共设施的投资，该项目对佩斯维尔市当前面对的主要城市挑战做出了回应。

该地块包含一幢1903年建造的历史建筑——老水星住宅的遗留结构，以及冷战时期建造的两个地下金库。老水星住宅破损的外立面已被修复，充满历史感的室内空间被保留下来并重新布置为通往公寓和酒店入口的公共空间。

这座31层高的塔楼包含住宅公寓和酒店两个部分，其平面布局与佩斯维尔现有的城市肌理相融合，以减少首层占地面积，并最大限度地增加新建城市广场内的市民空间。大厦的设计概念为两个垂直堆叠的体块，通过重新调整以容纳不同的内部功能。下面的9层为公寓，上面的19层为酒店，上部体块通过旋转使得酒店客房可以直接面向地中海。水星大厦的重建工作结合了公共、住宅和商业功能，打造了一个充满活力的城市空间，响应了该岛未来社会经济发展的需求，同时保护、改造了废弃的历史建筑。

Mercury Tower

Paceville, Malta
2016–2023
J. Portelli Projects

The renovation and redevelopment of Mercury Tower integrates residential apartments and a boutique hotel in the dynamic urban environment of Paceville at the heart of St. Julian's, Malta. With a rich history dating back to the early 20th century and renowned as a thriving tourist destination since the 1960s, Paceville boasts a multitude of restaurants, bars, nightclubs, casinos, marina and international hotels. Creating new public spaces and amenities for the island's residents and visitors, the design of the architecture responds to Paceville's key urban challenges by investing in its civic realm and increasing its limited housing supply.

The site incorporates the heritage structures of the old Mercury House that dates from 1903, as well as two underground vaults built during the Cold War. The old Mercury House façades have been restored, and the remaining historic interiors have been reinstated as the entrance and gathering spaces for the apartments and hotel.

The 31-storey tower, housing residential apartments and a hotel, has been carefully aligned and integrated into Paceville's existing urban fabric at street level, minimising its footprint and maximising civic space within the new piazza. The tower is conceived as two volumes stacked vertically, with a realignment that expresses the different functional programmes within. The lower 9-storey volume houses apartments, while the higher 19-storey volume is rotated to provide Mediterranean-facing guest rooms for the new hotel. By combining public, residential, and commercial functions and creating a vibrant civic space, the redevelopment of Mercury House responds to the demands of the island's future socio-economic development while preserving and renovating derelict heritage structures.

透视 01

2022年
ILLULIAN

透视01地毯是ZHA设计的"建筑系列"的一部分,专为意大利品牌ILLULIAN而设计,使用喜马拉雅山绵羊毛和丝线手工制作而成。ILLULIAN是一家历史悠久且享有盛誉的地毯公司,专门从事豪华手工地毯定制。该设计展现了ILLULIAN的精湛工艺与ZHA设计理念之间的巧妙融合,将透视、形式和空间完美地还原在地毯上。

"建筑系列"特点鲜明,大胆、前卫,富有张力和流动感。直线与曲线的穿插交织,通过地毯的材质变化进一步增强。有光泽的丝线与暗色的羊毛形成对比,创造出图案与背景之间的动态交融。这种交融的关系随着光线的起伏而变化,给人一种三维效果的错觉,同时挑战并重新定义了"空间"概念。

除"建筑系列"外,ZHA还设计了"自然领域"系列,均于2022年推出,反映了ZHA在建筑领域的拓展研究与专业性,以及他们对ILLULIAN地毯材料与工艺的探索和运用。

Perspective 01 Rug

2022
ILLULIAN

The Perspective 01 Rug is part of the Architectural Collection designed by Zaha Hadid Architects and handmade using Himalayan wools and silks. Designed for the Italian brand ILLULIAN, a historic and prestigious rug company specialising in luxury custom handmade rugs, it exemplifies the connection between ILLULIAN's craftsmanship and Zaha Hadid Architects' design philosophy, which seamlessly integrates perspective, form, and space.

The Architectural Collection is characterised by extreme dynamism and fluidity of the defined—sometimes daring—lines, which alternate with expressive curves that are intensified by the materiality of the rugs. Interwoven threads of silk produce a sheen that interacts with the matte of the wool, generating dynamic interplays between figure and ground. These interplays continually change with the light, giving the illusion of a three-dimensionality that both challenges and redefines the notion of space.

In addition to the Architectural Collection, ZHA developed the Natural Field Collection. Both collections were launched in 2022 and reflect ZHA's architectural research and expertise, as well as their exploration of ILLULIAN's rug-making materials and processes.

算法设计

扎哈·哈迪德建筑事务所（ZHA）的算法设计研究组（ZHA CODE）是一个实践嵌入的研究团队。ZHA CODE专注于实现从学科进步到实际应用的战略创新，坚持以未来为导向的设计价值观，同时从传统的智慧中汲取经验，尤其是在砖石和木材建筑、气候适应性、空间布局和细节等方面的乡土智慧。

Zaha Hadid Architects CODE

Zaha Hadid Architects Computation and Design Group (ZHA CODE) is a practise-embedded research group focussing on strategic innovations that bridge from disciplinary advances to novel and practical applications. The group's work, guided by their deeply held values of future-oriented design also learns from the wisdoms of the past—particularly local and vernacular wisdoms in masonry and timber construction, climatic appropriateness, and spatial arrangements and details.

以高性能为基础的美学

ZHA CODE 的设计是由最先进的、专有的参数化设计软件和技术驱动,并兼顾物理性能和社会性能的美学结果。研究组使用的工具经过 15 年以上的专业开发和实地测试,覆盖计算机图形学和几何学、数字设计、日趋成熟的机器人建造技术,以及传统的木材与砖石建造等领域。在社交互动层面,ZHA CODE 的设计得益于对参与式技术的研究和开发,使以用户为中心的设计成为可能——允许社交互动引导设计,并为非专业人士和最终用户提供属于专业领域的流程体验和访问路径。

以责任为导向的设计

ZHA CODE 的研发工作确保了 ZHA 的设计典雅、独特且面向未来。此外,不断迭代的技术还确保了 ZHA 流体、动态的几何设计在结构方面有可靠的理论和实践依据,且重视生产需求、符合环保理念。这就意味着,ZHA 的设计能够真正达到高性能的要求:一方面实现低碳排放、低能耗和轻质化;另一方面又能保持高强度和耐久性。ZHA CODE 的理念、项目、技术和合作伙伴们旨在打造适应 21 世纪的建筑和城市规划——交互密集、体验丰富、以用户为中心、资源高效。

Aesthetics of High-Performance

The aesthetics of designs at ZHA CODE is a consequence of its physical and social high-performance. They are powered by state-of-the-art, proprietary parametric design software and technologies. These tools are a result of over 15 years of dedicated and field-ready research spanning computer graphics and geometry, digital design, maturing robotic construction technologies, historical construction in timber and masonry, etc. On the social side, these designs benefit from research and development in participatory technologies enabling user-focussed design, social interactions to inform design and provide access to typically expert-domain processes to non-experts and end-users.

Responsible Design

The research and development at ZHA CODE ensure that the designs of ZHA are elegant, distinctive, and future-ready. Furthermore, the ever-evolving technologies also ensure that the fluid, dynamic geometries associated with ZHA, are structurally informed, production-aware, and environmentally appropriate. This means that the shapes can truly balance high-performance requirements in terms of being low-carbon emissions, low-embodied-and-operational energy, and lightweight on the one hand, whilst having high-strength and endurance on the other hand. Together, the group's ethos, projects, technologies, and collaborators aim to create an architecture and urbanism suited for the 21st century—interaction dense, experience rich, user-focussed, and resource effective.

Striatus 3D 混凝土打印砌块人行拱桥，威尼斯双年展，意大利，威尼斯，马尔纳雷萨花园，2021 年
Striatus 3D Concrete Printed Bridge, Venice Biennale, Giardini della Marinaressa, Venice, Italy, 2021

研究轨迹简介

ZHA CODE探索新技术在结构、可建造的几何形状、机器人制造和工业化建造等领域所带来的全新空间和体验价值。研究组将这些进一步融入ZHA的工作流程中，以解决建筑环境中参与式和可持续发展的紧迫问题。ZHA CODE研究属于设计领域的问题，用计算机设计去补充人类的直觉、提高速度、整合数据，并发现新的可能性。

研究组通过系统发展和对关注轨迹的试点测试逐步积累基础研究主题。随着时间的推移，这些主题不断得到完善，方法也日趋成熟并逐渐适应商业项目应用环境下的限制。由此发展出的定制计算框架继续为合作研究和实地测试做积累，多样化的主题涵盖计算机图形学、机器人建造技术，等等。这些工具、方法和工作流程承载着数学、物理和材料的历史，并在当代环境中继续使用，形成了一个连续的进程。当前的研究涵盖计算机图形学、几何处理技术、制造技术、数据驱动设计、游戏技术、一体式交付和供应链整合。

本次展览中着重展示了以下五个研究轨迹：
1. 机器人辅助设计
2. 规则化和可展开结构
3. 数字木构
4. 极小曲面
5. 图形及函数的表达

Introduction to Research Trajectories

ZHA CODE explores novel spatial and experiential affordances of new technologies in structure, fabrication-aware geometry, robotic fabrication, and industrialised construction. The group incorporates these advances into workflows at ZHA to address pressing issues of participatory and sustainable development of the built environment. The group investigates design domain problems where computation can supplement human intuition, to improve speed, to assimilate data, and to discover novel possibilities.

Research is incrementally accrued through systemic development and pilot testing in fundamental topics or trajectories of focus. These are refined over time, and methods mature toward increasingly constrained commercial project applications. The resulting bespoke computation framework continues to accrue collaborative and field-tested research in topics ranging from computer graphics to robotic construction technologies. These tools, methods, and workflows inherit the mathematical, physical, and material history, and employ them in a contemporary setting as a continuum. Current research ranges across Computer Graphics, Geometry Processing, Fabrication Technologies, Data Driven Design, Game Technologies, Package Delivery, and Supply Chain Integration.

This exhibition focuses on these five research trajectories:
1. Robotic Assisted Design
2. Ruled and Developable Structures
3. Digital Timber
4. Minimal Surfaces
5. Graph and Function Representation

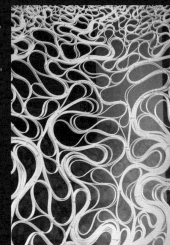

1. 机器人辅助设计
Thallus，米兰设计周，意大利，米兰，2017年
1. Robotic Assisted Design
Thallus, Milan Design Week, Milan, Italy, 2017

2. 规则化和可展开结构
第聂伯罗地铁站，乌克兰，第聂伯罗，2023年（进行中）
2. Ruled and Developable Structures
Dnipro Metro Stations, Dnipro, Ukraine, 2023 (ongoing)

3. 数字木构
罗阿坦·普罗斯佩拉住宅，洪都拉斯，罗阿坦岛，2022年
3. Digital Timber
Roatán Próspera Residences, Roatán Island, Honduras, 2022

4. 极小曲面
数学：温顿画廊，科学博物馆，英国，伦敦，2014—2016年
4. Minimal Surfaces
Mathematics: The Winton Gallery, Science Museum, London, UK, 2014–2016

5. 图形及函数的表达
Striatus 3D混凝土打印砌块人行拱桥，2021年
5. Graph and Function Representation
Striatus 3D Concrete Printed Bridge, 2021

1. 机器人辅助设计

结合现有生产过程中制造和建造局限性的设计流程正在得到越来越多的重视。ZHA CODE 的研究探讨了在设计建模环境中为6轴工业机器人集成逆向运动（IK）链条的好处。这有助于在设计模型和制造限制之间实现即时反馈，允许为机器人制造去改造设计的几何形状。此外，该工具能自动生成指定生产流程（如机器人热线切割、3D混凝土打印、纤维缠制等）的工具路径和格式化的机器G代码。

　　本次展览中展示了一系列机器人设计应用，如Striatus混凝土打印砌块人行拱桥和以数学为基础的温顿画廊的展览设计。

1. Robotic Assisted Design

Design pipelines that incorporate fabrication and construction constraints of existing manufacturing processes are increasingly valued. The research explores the benefits of integrating an Inverse Kinematic (IK) chain for 6-Axis industrial robots in a design modelling environment. This facilitates immediate feedback between design model and fabrication constraints and allows for making a design geometry amenable for robotic fabrication. Additionally, the tool enables automated generation of toolpaths and formatted machine g-codes for a given production pipeline such as robotic hot wire cutting, 3D concrete printing, fibre winding, etc.

　　In this exhibition, a range of robotic design applications are presented, such as the Striatus 3D Printed Concrete Bridge, and the permanent exhibition design for Mathematics: The Winton Gallery.

01　Thallus，米兰设计周，意大利，米兰，2017年
　　Thallus, Milan Design Week, Milan, Italy, 2017

02　坎德拉织物研究模型，当代艺术大学博物馆（MUAC），墨西哥，墨西哥城，2018—2019年
　　Study Models for Knit Candela, MUAC, Mexico City, Mexico, 2018-2019

03　长椅研究模型，数学：温顿画廊，科学博物馆，英国，伦敦，2014—2016年
　　Bench Study Models, Mathematics: The Winton Gallery, Science Museum, London, UK, 2014-2016

04　3D打印椅子研究模型，2016年
　　Study Models for 3D Printed Chair, 2016

05　Striatus 3D混凝土打印砌块人行拱桥展示模型，2021年
　　Presentation Model for Striatus 3D Printed Concrete Bridge, 2021

06　Cirratus 3D打印混凝土花瓶，2017年
　　Cirratus 3D Printed Concrete Vase, 2017

07　Striatus 3D混凝土打印砌块人行拱桥展示模型，2021年
　　Presentation Model for Striatus 3D Printed Concrete Bridge, 2021

08　织物形体研究模型，数学：温顿画廊，科学博物馆，英国，伦敦，2014—2016年

2. 规则化和可展开结构

数学理论下的规则化几何表面已诞生数百年，但对它们的研究在微积分发明后才开始深入。这些几何表面在19世纪的砌体和木材结构中非常突出并被广泛使用，最近也用于使复杂的建筑几何结构合理化。可展开的几何图形是一组特殊的几何图形：除了每个面都是平面的，它们还可以在没有畸变的前提下展开到平板上。这种表达使得几何图形适合用相对便宜的2轴CNC切割技术进行制造。以大卫·赫夫曼、罗恩·雷施以及如理查德·斯威尼等折纸艺术家的先前工作为基础，弯曲褶皱折叠（CCF）几何形态的数字设计和建筑应用开始得到广泛关注。ZHA CODE 的研究强调通过对替代材料（如纸张）的物理探索来进行设计，以实现对金属、塑料和纤维复合材料的使用。

2. Ruled and Developable Structures

Mathematically ruled surfaces have been known for centuries, but the in-depth study of them gained traction only after the invention of calculus. These surfaces were very prominent and widely used in 19th-century masonry and timber structures, and more recently to rationalise complex architectural geometries. Developable geometries are a special set of geometries: in addition to each face being planar, these can be unrolled to flat sheet without distortion. Such a representation makes the geometry amenable for fabrication via relatively inexpensive 2-axis CNC cutting technology. Building on the historic work of David Huffman, Ron Resch, and paper artists such as Richard Sweeney, there has been increasing interest in the digital design and architectural application of Curved Crease Folded (CCF) geometries. The ZHA CODE research emphasises design through physical exploration, in proxy materials such as paper, for realisation in metal, plastics, and fibre composites.

01　西安国际足球中心的计算几何研究，中国，西安，2020—2023年
Computational Geometry Study for Xi'an International Football Stadium, Xi'an, China, 2020–2023

02　曲线褶皱折叠摇椅原型，2012年
Prototypes for Curve Crease Folded Rocking Chair, 2012

03　曲线褶皱折叠穿孔托盘研究模型，2012年
Study Model for Curve Crease Folded Perforated Serving Platter, 2012

04　金融大楼研究模型，利伯兰元宇宙，2021年（进行中）
Study Model for Finance Building, Liberland Metaverse, 2021 (ongoing)

05　研究模型，独角兽岛城市设计梯度迷你塔楼，中国，成都，2018年（进行中）
Study Models, Gradient Mini Towers for Unicorn Island Masterplan, Chengdu, China, 2018 (ongoing)

06　彩色幕墙剖面研究模型，独角兽岛城市设计，中国，成都，2018年（进行中）
Colour Façade Section Study Model, Unicorn Island Masterplan, Chengdu, China, 2018 (ongoing)

07　第聂伯罗地铁站研究模型，乌克兰，第聂伯罗，2018年（进行中）
Study Model for Dnipro Metro Stations, Dnipro, Ukraine, 2018 (ongoing)

08　沐梵世度假酒店皮耶·艾曼休息室的计算几何研究，中国，澳门，2022年
Computational Geometry Studies for Morpheus Pierre Herme Lounge, Macau, China, 2022

09　坎德拉织物原型，当代艺术大学博物馆（MUAC），墨西哥，墨西哥城，2018—2019年
Prototype for Knit Candela, MUAC, Mexico City, Mexico, 2018–2019

10　曲线褶皱折叠百褶顶棚提案研究模型，2018年
Study Model for Curve Crease Folded Pleated Canopy Proposal, 2018

11　Volu用餐亭原型，与扎哈·哈迪德设计（ZHD）合作，2015年
Prototype for Volu Dining Pavilion, in Collaboration with Zaha Hadid Design, 2015

12　金融大楼研究模型，利伯兰元宇宙，2021年（进行中）
Study Model for Finance Building, Liberland Metaverse, 2021 (ongoing)

13　3D打印张力椅研究模型，2016年
Study Models for 3D Printed Tensile Chair, 2016

14　第聂伯罗地铁站壳体概念研究模型，乌克兰，第聂伯罗，2018年（进行中）
Concept Shell Study Model for Dnipro Metro Stations, Dnipro, Ukraine, 2018 (ongoing)

15　第聂伯罗地铁站壳体概念研究模型，乌克兰，第聂伯罗，2018年（进行中）
Concept Shell Study Model for Dnipro Metro Stations, Dnipro, Ukraine, 2018 (ongoing)

16　沐梵世度假酒店皮耶·艾曼休息室的计算几何研究，中国，澳门，2022年
Computational Geometry Study for Morpheus Pierre Hermé Lounge, Macau, China, 2022

17　参数曲面类型研究，2016年
Parametric Surfaces of Various Typologies, 2016

18　展馆研究模型，木结构屋顶剖面，独角兽岛城市设计，中国，成都，2018年（进行中）
Study Model of Pavilion, Timber Structure Roof Section, Unicorn Island Masterplan, Chengdu, China, 2018 (ongoing)

19　Volu用餐亭研究模型，与扎哈·哈迪德设计（ZHD）合作，2015年
Study Models for Volu Dining Pavilion, in Collaboration with Zaha Hadid Design, 2015

20　平台开发研究模型，2022年
Study Models for Platform Technology, 2022

3. 数字木构

该研究探索了可持续轻型木材和工程木制品在数字化建造和制造中的应用。这种低碳且省力的解决方案得益于对当地木料的使用和模块化的思维方式。异地制造（OSF）的理念减少了材料的浪费并且提高了施工的精度和质量。解决方案以"及时"这一概念为基础，即工厂通过协调和交付"零件套件"的方式去实现在现场的快速组装。木结构设计和木构件的布置被广泛地探索，包括单曲板、直短组件的堆叠、平面轮廓设计、胶合木和双曲组件。

ZHA从现代家具设计和传统细木工及手工艺技术中汲取灵感，对单曲木材产品使用塑形、弯曲和单板叠层等工艺技术，旨在通过采用工业化机器人进行数字化设计、建造和组装。

同样，在单层结构中使用直短组件，并且在生成有限网格时，将组件与找形相结合，由此发展出在大跨中应用的双层桁架网络和混合悬链线系统。

3. Digital Timber

The research explores the use of sustainable lightweight timber and engineered timber products in digitised construction and manufacturing. By incorporating locally sourced timber and adopting modular thinking, the research aims to create solutions that effectively minimize the embedded construction energy and carbon footprint of the schemes. Off-site fabrication allows for a reduction in waste material and ensures a higher quality of construction. The schemes are designed around the 'just-in-time' concept, wherein the factory coordinates the delivery of modules as a 'kit of parts,' allowing for swift assembly on-site. Timber design and element setout are broadly explored, encompassing singly curved sheets, short linear elements in aggregate, planar profiles, glulams, and doubly curved members.

The generation of geometries in singly curved timber products speculates on the use of forming, bending, and lamination of sheet veneers, drawing inspiration from both modern furniture design and historic joinery and handcraft techniques. This approach aims to advance digital design, construction, and assembly through the utilisation of industrialised robotics.

Likewise, when element setout is coupled with form finding in the generation of finite element networks, short linear elements are used with variations in joints for single-layer structures. Subsequently, these methods have been further developed for long-span applications, including dual-layer truss networks and hybrid catenary systems.

01 空间弯曲层压木结构，2021年
Spatial Curved Laminated Timber Structure, 2021

02 平台开发——住宅配置模拟，罗阿坦·普罗斯佩拉住宅，洪都拉斯，罗阿坦岛，2020年（进行中）
Study Model for Platform Technology - Housing Configurator, for Roatán Próspera Residences, Roatán Island, Honduras, 2020 (Ongoing)

03 空间弯曲层压木结构，2021年
Spatial Curved Laminated Timber Structure, 2021

04 空间弯曲层压木结构，2021年
Spatial Curved Laminated Timber Structure, 2021

05 罗阿坦·普罗斯佩拉住宅的形体研究模型，洪都拉斯，罗阿坦岛，2020年（进行中）
Massing Study Model for Roatán Próspera Residences, Roatán Island, Honduras, 2020 (Ongoing)

06 罗阿坦·普罗斯佩拉住宅的形体研究模型，洪都拉斯，罗阿坦岛，2020年（进行中）
Massing Study Model for Roatán Próspera Residences, Roatán Island, Honduras, 2020 (Ongoing)

07 结构模型研究，动荡时刻，士瑞克保全公司，2018年
Study Model for Disruption Days G4S Secure Solutions, 2018

08 豪华别墅研究模型，2022年
Study Model for Luxury Villas, 2022

09 国会大厅研究模型，利伯兰元宇宙，2021年
Study Model for Congress Hall, Liberland Metaverse, 2021

10 维也纳竞技场数字木构研究，奥地利，维也纳，2021年
Digital Timber Study for Vienna Arena, Vienna, Austria, 2021

11 楼板研究模型，2022年
Floor Slab Study Model, 2022

12 维也纳竞技场数字木构研究，奥地利，维也纳，2021年
Digital Timber Study for Vienna Arena, Vienna, Austria, 2021

13 楼板研究模型，2022年
Floor Slab Study Model, 2022

14 Roatán Próspera住宅基于木材的建筑体量和数字化研究，洪都拉斯，罗阿坦岛，2020年（进行中）
Volumetric and Digital Timber Studies for Roatán Próspera Residences, Roatán Island, Honduras, 2020 (Ongoing)

15 结构模型研究，动荡时刻，士瑞克保全公司，2018年
Study Model for Disruption Days G4S Secure Solutions, 2018

4. 极小曲面

极小曲面是一种类似于肥皂膜的特殊几何体，由局部平均曲率趋向于零的表面组成。弗雷·奥托在给定边界的情况下通过对肥皂膜的物理找形来实现对该几何体的研究。这类几何体在数学领域中也曾被探索[1]。这些表面的计算过程也被称为"找形"，通常采用两种主流方法，即动态松弛法[2]和力密度法[3]。

包括ZHA（CODE）在内的多家建筑和工程公司都对可拉伸索以及纺织形体等建筑实践进行了研究。ZHA在伦敦科学博物馆"数学：温顿画廊"（2014—2016年）项目中使用的就是动态松弛法，它最初开发于1965年，用于寻找在轴向力作用下的几何体的平均形态。

4. Minimal Surfaces

Minimal surfaces are a special geometric set, similar to soap films, comprised of surfaces tending toward zero mean curvature locally. Frei Otto studied them via physical form-finding of soap films that form against a given wire boundary. These geometries have also been studied mathematically.[1] The computational generation of these surfaces, known as a form-finding process, typically involves one of two popular methods: the Dynamic Relaxation Method[2] and the Force Density Method[3].

The architectural materialisation of these methods as stretched cable and fabric forms has been studied by several architectural and engineering firms, including the ZHA CODE group. Mathematics: The Winton Gallery, designed by Zaha Hadid Architects for the Science Museum in London (2014–2016), utilises Dynamic Relaxation, a computational method originally developed by Day to find equilibrium shapes of geometries subjected to axial forces.

---

1. Brakke K A. 极小曲面，角点和线 [J]. 几何分析杂志, 1992, 2(1): 11-36.
2. Day A S. 动态松弛概论 [J]. 英国工程师杂志, 1965, 219: 218-221.
3. Schek H.-J. 力密度法用于找形和一般网络计算 [J]. 瑞士应用力学与工程中的计算方法杂志, 1974, 3(1): 115-134.

1. Brakke, K. A. (1992). Minimal surfaces, corners, and wires. *The Journal of Geometric Analysis*, 2(1), 11–36.
2. Day, A. S. (1965). An introduction to dynamic relaxation. *The Engineer*, 219, 218–221.
3. Schek, H.-J. (1974). The force density method for form finding and computation of General Networks. *Computer Methods in Applied Mechanics and Engineering*, 3(1), 115–134.

---

01 交互式装置，"扎哈·哈迪德：世界建筑系列"，丹麦建筑中心，丹麦，哥本哈根，2013年
Interactive Installation, "Zaha Hadid: World Architecture Series" at Danish Architecture Centre, Copenhagen, Denmark, 2013

02 坎德拉织物研究模型，当代艺术大学博物馆（MUAC），墨西哥，墨西哥城，2018—2019年
Study Model for Knit Candela, MUAC, Mexico City, Mexico, 2018–2019

03 坎德拉织物研究模型，当代艺术大学博物馆（MUAC），墨西哥，墨西哥城，2018—2019年
Study Models for Knit Candela, MUAC, Mexico City, Mexico, 2018–2019

04 织物找形下的混凝土壳体结构，建筑联盟学院班加罗尔访校，印度，班加罗尔，2011年
Fabric Guide-work Concrete Shell, AA Visiting School, Bangalore, India, 2011

05 研究模型，数学：温顿画廊，科学博物馆，英国，伦敦，2014—2016年
Study Models for Mathematics: The Winton Gallery, Science Museum, London, UK, 2014–2016

06 壳体研究模型，班加罗尔建筑联盟学院访学，印度，班加罗尔，2011年
Shell Study Model for AA Visiting School, Bangalore, India, 2011

07 西安国际足球中心的计算几何研究，中国，西安，2020—2023年
Computational Geometry Study for Xi'an International Football Stadium, Xi'an, China, 2020–2023

08 参数曲面类型研究，2016年
Parametric Surfaces of Various Typologies, 2016

09 织物形体研究模型，数学：温顿画廊，科学博物馆，英国，伦敦，2014—2016年
Fabric Pod Study Model for Mathematics: The Winton Gallery, Science Museum, London, UK, 2014–2016

10 难民学校研究模型，2015年
Study Model for Refugee School, 2015

11 数学：温顿画廊，科学博物馆，英国，伦敦，2014—2016年
Mathematics: The Winton Gallery, Science Museum, London, UK, 2014–2016

5. 图形及函数的表达

建筑几何体的表示方法通常分为：
- 显式表达，包括点云、图形、多边形网格、NURBS 曲面等；
- 隐式表达，包括数值表达、解析表达、函数表达等。

图形，是一种由点和线组成的轻量级的几何表达方法。在建筑的语境下，图形通常以骨架图的形式去表示楼板的中心线或者结构推力网格，它可以作为生成下游几何体的基础，也可以作为生成标量有向距离场的输入值。

函数表达（FRep）通常被用于实体建模和计算机图形学，并且常被可视化为按比例伸缩的标量图像或者一系列二维、三维场轮廓[1]。Striatus 3D 混凝土打印砌块人行拱桥使用图形轮廓以及渐变的距离场去创建 3D 打印的节点，并通过三维图形静力学的方法去找形骨格线。

在 2021 年威尼斯双年展期间，Striatus 首次在 Marinaressa 花园中组装，该项目由苏黎世联邦理工学院的 Block 研究小组（BRG）、ZHA CODE、incremental3d（in3D）共同开发，并由 Holcim 公司提供支持。这座 16 × 12 米的人行拱桥由 3D 打印混凝土砌块组成，无须砂浆或钢筋即可实现组装，为传统混凝土建筑提供了一种替代方式。

5. Graph and Function Representation

Architectural Geometry representations are broadly classified into:
• Explicit representations, which include point clouds, graphs, polygonal meshes, NURBS, etc.
• Implicit representations which include numerical, analytical, function representation, etc.

Graphs are one of the lightweight geometrical representations, as they consist of a collection of points and lines. In architectural scenarios, they are commonly used as skeletal diagrams representing the centre line of a floor plate or structural thrust networks. These graphs serve as a basis for procedurally generating downstream geometries or as inputs for generating scalar signed distance fields.

Function Representation (FRep) is typically used in solid/volume modelling and computer graphics. FRep was introduced by Alexander Pasko et al. (1995), and is usually visualised as scalar field image stacks or 2D / 3D contours of the fields[1]. Striatus, the 3D Concrete-Printed Masonry Footbridge, is a project that uses the contours and blends of a sign distance field of a skeletal graph to create a 3D printed block. The skeletal graph was form-found using Graphic Statics.

First assembled in the Giardini della Marinaressa, on occasion of the 2021 Venice Biennale, Striatus was developed by the Block Research Group at ETH Zurich and ZHA CODE in collaboration with incremental3d and made possible by Holcim. The 16 x 12 metre footbridge is composed of 3D-printed concrete blocks assembled without mortar or reinforcement, presenting an alternative to traditional concrete construction.

01　计算几何研究，沐梵世度假酒店皮耶·艾曼休息室，中国，澳门，2022 年
Computational Geometry Study for Morpheus Pierre Herme Lounge, Macau, China, 2022

02　空间弯曲层压木结构，2021 年
Spatial Curved Laminated Timber Structure, 2021

03　结构节点研究模型：锯齿状木质模板与浇铸，2018 年
Study Model for Structural Node: Zip Shape Formwork and Cast, 2018

04　3D 打印结构节点的研究模型，2021 年
Study model for 3D printed structural Node, 2021

05　Striatus 3D 打印混凝土桥展示模型，2021 年
Presentation Model for Striatus 3D Printed Concrete Bridge, 2021

06　木制结构的 3D 打印节点，1∶50 比例，2019 年
3D Printed Node for a Timber Structure, 1∶50 (Scale) 2019

07　空间弯曲层压木结构，2021 年
Spatial Curved Laminated Timber Structure, 2021

08　结构节点研究模型：金属折叠和接缝，2016 年
Study Model for Structural Node, Folded and Seamed, 2016

09　第聂伯罗地铁站研究模型，乌克兰，第聂伯罗，2018 年（进行中）
Study Model for Dnipro Metro Stations, Dnipro, Ukraine, 2018 (ongoing)

10　香港科技大学学生公寓项目初步研究模型，中国，香港，2018 年
Preliminary Study Model of Student Residence Development at Hong Kong University of Science and Technology, Hong Kong, China, 2018

11　Striatus 3D 打印混凝土桥展示模型，2021 年
Presentation Model for Striatus 3D Printed Concrete Bridge, 2021

---

1. Pasko A, Adzhiev V, Sourin A, Savchenko V. 几何建模中的函数表达：概念、实现和应用 [J]. 可视计算机期刊，1995, 11(8): 429-446.

1. Pasko, A., Adzhiev, V., Sourin, A., & Savchenko, V. (1995). Function representation in geometric modeling: Concepts, implementation and applications. *The Visual Computer, 11*(8), 429–446.

机器人建造

机器人建造延续了ZHA CODE对建筑几何学（AG）的研究——一种专注于创造优化结构与制造的形状的设计技术范式。

AG结合大型3D打印技术，为此次展览的机器人建造展区打造了一个名为"玉屏"的11米长的空间界面。它由63个高度在1.2~2米的离散体块组成。这些体块将由机械臂现场打印，并在展览期间现场组装。该装置探索了计算机图形学中常用的计算技术如热传导方法、标量场和等值切面法，生成适合机器人3D打印的几何图形与构造。

抽象地图"Diffusa"是采用现代空间填充曲线与差分增长算法设计而成的，并结合了城市与城市数据集，如街道网络、空地与绿地、人口密度、街区建筑高度等。从基本街道网格开始，这些数据集被用于控制曲线图案密度、城市街区的高度与颜色等模拟参数。79个城市街区分别用曲线填满，然后通过挤压、拉伸来创建3D打印图案。

展览期间，两个装置由机器人现场打印并组装而成。北京建筑大学协助ZHA利用上海大界机器人科技有限公司提供的机械臂进行3D打印。

Robotic Assisted Design

The Robotic Assisted Design continues Zaha Hadid Architects Computation and Design Group's (ZHA CODE) investigations into Architectural Geometry (AG)—a design technology paradigm that focuses on creating shapes that are optimised for structure and fabrication.

AG in combination with large-scale 3D printing technology is employed to create an 11-metre spatial screen, titled Yuping, for the Robotic Assisted Design zone of the gallery. It is composed of 63 discrete blocks with variable height (1.2–2 metres). These will be printed in-situ by the robotic arm and assembled in the gallery over the course of the exhibition. The installation explores the usage of computational techniques typically used in computer graphics—heat method, scalar-fields, and iso-slicing—to generate geometries and tectonics amenable to robotic 3D printing.

The design of the abstract Beijing map, Diffusa, employs contemporary space filling curves and differential growth algorithms in combination with the city and urban datasets such as street networks, open and green spaces, population density, block building heights, etc. Starting out with the base street grid, the datasets are used to control the simulation parameters of squiggle pattern density, the height and colour of the city blocks. A squiggle curve fills up each of the 79 city blocks, which are then variably extruded to create a 3D printed tile. The tiles are robotically printed and assembled during the exhibition.

The Beijing University of Civil Engineering and Architecture oversaw the 3D printing of both the designs, on the robotic arm supplied by Robotic Plus, Shanghai.

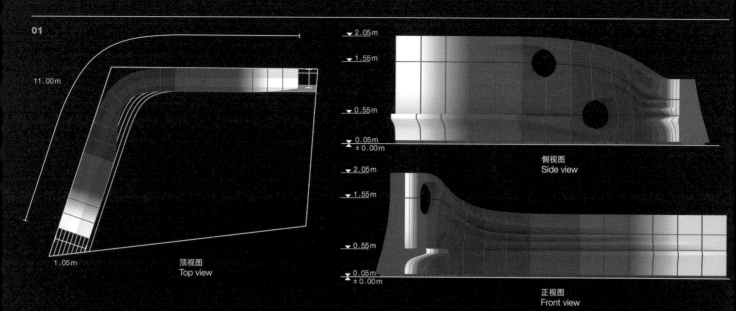

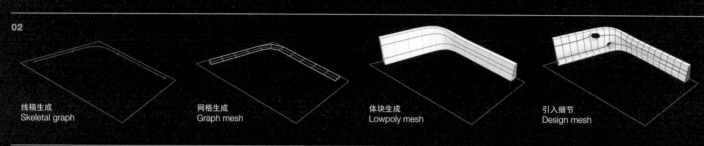

01 概念图纸	02 设计流程	03 展览中正在进行组装的过程照片
Schematic design	Procedural design workflow	Photo of the in progress assembly in the gallery

16.68	822.8	185.4	63
打印长度（千米）	重量（公斤）	打印时长（小时）	体块数量
Print length (km)	Weight (kg)	Print time (h)	Number of block

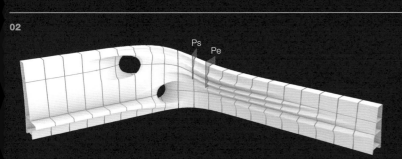
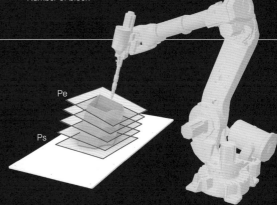
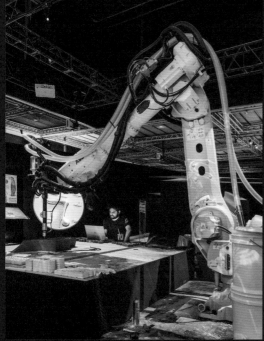

01   3D打印模块透视图
Perspective view showing discrete blocks for 3D printing

02   显示打印起始平面和终止平面（Ps, Pe）的曲线3D打印模块，利用了非平面打印技术
Curved 3D printed blocks showing the start and end plane (Ps, Pe), which are produced using multi-planar robotic 3D printing

03   展览中的机器人3D打印
In-situ robotic 3D printing in the exhibition

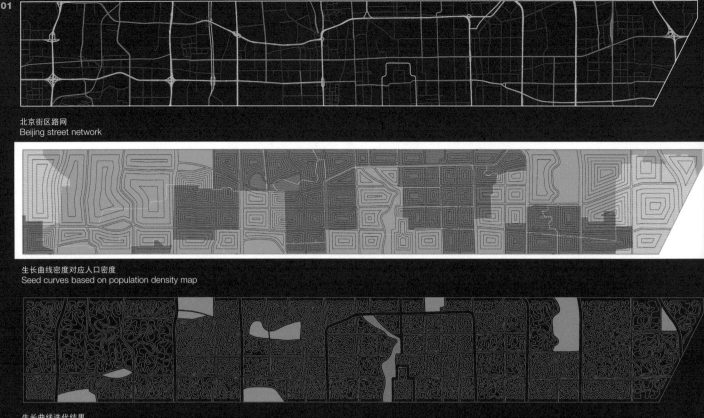

北京街区路网
Beijing street network

生长曲线密度对应人口密度
Seed curves based on population density map

生长曲线迭代结果
Output pattern

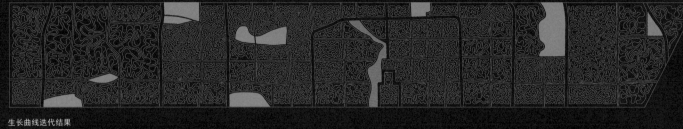

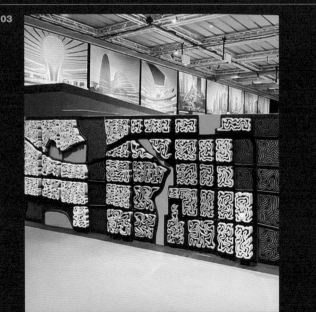

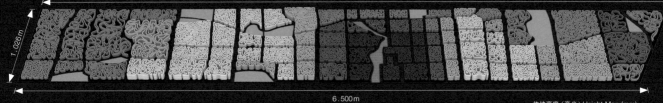

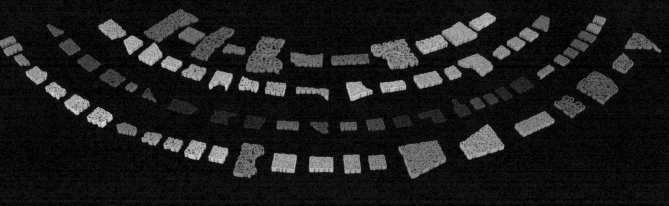

2.79	137.4	30.9	80
打印长度（千米）	重量（公斤）	打印时长（小时）	体块数量
Print length (km)	Weigh (kg)	Print time (h)	Num of block

01　3D打印模块透视图
　　Perspective view showing discrete blocks for 3D printing

02　效果图
　　Rendered view

数字社会

扎哈·哈迪德建筑事务所（ZHA）的数字社会研究组（ZHA Social）致力于开发一种新的研究方法来比较不同设计方案的社会性能。该研究的基本假设是人类行为与建筑空间是相互关联的。我们可以通过模拟与测量这种关系来预测所设计空间的社会性能。ZHA Social 通过多智能体系统来研究建筑环境中预期的社会互动过程，从而研究人类生活场景与建筑之间的动态关系，以便对社交行为进行主动的度量与预测。研究主要侧重于设计具有社交有效性的办公空间，同时对虚拟环境及事件中的社交互动进行拓展。

  这项研究包括三个平行开发的组成部分。其一，多智能体计算系统框架，可用于模拟与建筑空间规划相关的群体社会行为。个体模型根据设计中的建筑环境做出决策，生成群体空间占用数据。其二，具有深度学习能力的生成式空间规划模型，可根据特定的人力状况，快速生成与之相适应的办公空间设计，并通过模拟对社交与空间的性能进行分析，以增强协作。其三，沉浸式多用户体验模拟器，允许用户体验每个设计作品，与之互动并参与评估，同时基于虚拟模型对用户行为进行研究。这三个部分被整合到 Workplaces.AI 平台上，以促进数据驱动的协作设计过程。

Zaha Hadid Architects Social

The Zaha Hadid Architects Social Research Group (ZHA Social) focuses on the development of a new methodology for comparing the social functionality of design options. The underlying hypothesis of this research is that human behaviour is interrelated with architectural space, and this relationship can be simulated and measured to predict the social performance of the spaces we design. The group investigates the social interaction processes to be expected within architectural environments through agent-based simulations, enabling the study of the dynamic relationship between human life scenarios and architecture for the active measurement and prediction of social behaviour. The research primarily focuses on designing socially effective workspaces and extends to social interactions in virtual environments and events.

The research involves three main components developed in parallel. Firstly, an agent-based computational framework simulates social crowd behaviour in relation to architectural space planning. Individuals are modelled as agents making decisions in relation to designed architectural environments, generating collective occupancy Data. Secondly, a generative space planning model with deep learning rapidly designs adaptive workplaces tailored to the specific profile of the workforce, considering social and spatial performance criteria analysed through simulations to enhance collaboration. Thirdly, an immersive multi-user experiential simulator allows real users to experience, interact with, and evaluate each design while studying their behaviour within virtual models. These three components are integrated into a platform called Workplaces.AI, facilitating a collaborative data-driven design process.

Workplaces.AI 的应用

在展览中，ZHA Social 现场演示了 Workplaces.AI 平台，将其投影在广州无限极广场项目的物理模型上。该平台将研究组研发的三个主要组成部分结合在一个多用户平台上。这个集成了人工智能的生成设计工具可以快速生成数百个高性能的工作场所空间规划方案。多个用户可以实时协作，输入与空间组织相关的设计约束条件和目标，并选择工作场所的设计特征。用户与设计过程的互动促进了空间规划方案的生成。该平台用算法模拟人类社交行为，通过基于设计方案的模拟来分析一系列空间性能指标。此外，实时交互式体验模拟器能够让参与者对设计方案进行测试，并提供来自真实用户与参与者的附加数据。在模拟分析的基础上利用深度学习训练模型，用户能够快速生成和评估符合相关利益需求的设计方案。

Workplaces.AI Application

In the exhibition, ZHA Social showcases a live demonstrator of their platform, titled 'Workplaces.AI,' projected over a physical model of the Infinitus HQ project. Workplaces.AI combines the three main components of ZHA Social's research and development in a multi-user platform. The generative design tool with artificial intelligence rapidly generates hundreds of high-performance space planning options for workplaces. Multiple users can collaborate in real-time, inputting design constraints and goals related to spatial organisation and selecting workplace design features. The users' interaction with the design process drives the generation of space planning options. The platform utilises algorithms for simulating human social behaviour and a range of spatial performance criteria analysed through simulations in relation to each design option. Additionally, a real-time interactive experiential simulator enables stakeholders to test design options, providing additional data from real users and stakeholders in the process. Deep learning is leveraged to train models with simulation-based analytics, allowing users to quickly generate and evaluate design options that align with stakeholder requirements.

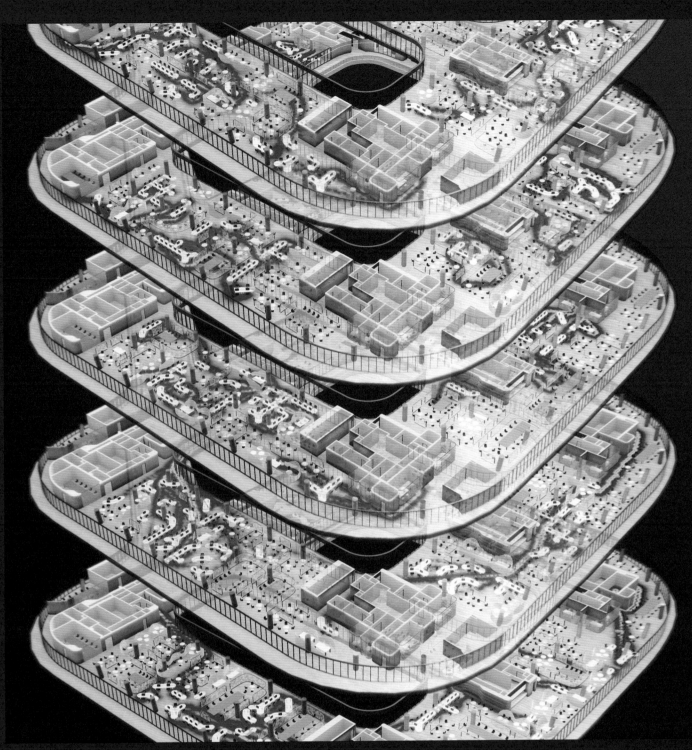

Workplaces.AI 生成方案，广州无限极广场，中国，广州
Workplaces.AI Generative Design for Guangzhou Infinitus Plaza, Guangzhou, China

展览数字孪生

ZHA Social 开发了云端多用户体验模拟工具及虚拟环境互动工具作为 Workplaces.AI 平台的一部分。ZHA-Meta-Workplaces 协作工具允许用户从任意地点以数字化身份登录，让他们能够实时加载并共同探索三维设计。用户可以通过沉浸式开放游览、草绘或笔记、屏幕共享与实时语音或视频等功能进行协作。在此次展览中，ZHA 扩展了这一功能，创造了沉浸式的展览数字孪生。展览数字孪生增加了信息层次，使用户能够在虚拟空间中穿行，与各种展览模型互动，并以 1:1 的真实比例自由游览与探索，从而增强了本次展览的互动体验。

Exhibition Digital Twin

As part of the Workplaces.AI platform, ZHA Social has developed tools for cloud-based multi-user experiential simulation and interaction with virtual environments. The ZHA-Meta-Workplaces collaboration tool enables groups of users to sign on as avatars from anywhere in the world, allowing them to load and explore 3D designs together in real-time. Users can collaborate through immersive open navigation, sketching/note taking, screen sharing, and live voice/video features. For this exhibition, the team has extended this functionality to create an immersive Exhibition Digital Twin. The Exhibition Digital Twin enhances the interactive experience of this exhibition by adding further information layers and enabling users to walk through the virtual space, interact with various exhibition models, and freely navigate and explore them at a 1:1 scale.

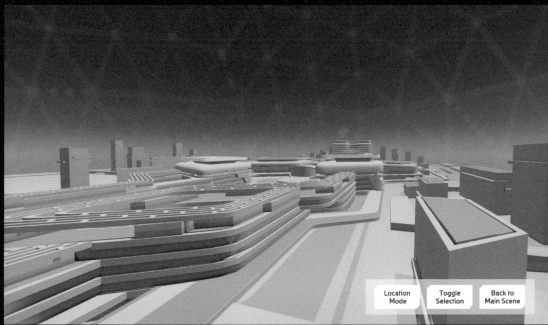
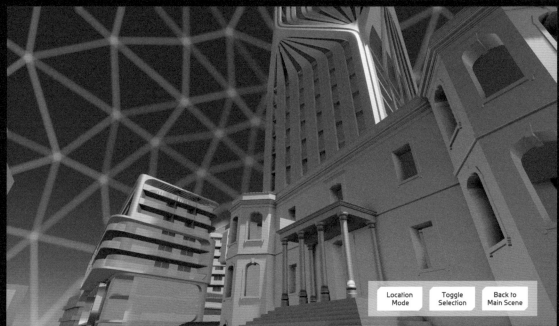

展览数字孪生屏幕截图
Screen captures from the Exhibition Digital Twin

可持续发展 ⑤

气候危机与碳消耗是全球普遍关注的问题，已超出人类的可控范围，这就迫使人们在建筑环境可影响的体系内采取紧急补救措施。扎哈·哈迪德建筑事务所（ZHA）的可持续性设计承诺强调建筑不仅是人类的栖身之所，更应为人类社会持续发展提供最高的指导原则。

ZHA致力于不断改进工作方式，希望在解决全球变暖这一关键问题中发挥带头作用。为了确保每个新项目都能持续取得更好的成果，ZHA按照国际公认的目标对项目的碳使用情况进行评估——了解其对环境的影响、明确现行最优的实践方案、梳理需要进行重大调整的设计思路。

ZHA可持续研究组应用详细的环境分析与形体优化技术，最大限度地提高项目效率；而ZHA低碳策略研究组则更强调资源节约型设计，从而降低资源浪费和碳影响。ZHA倡导在减少建筑的隐含碳和运营环节中的碳排放方面进行知识共享、策略优化。

经过ZHA内部的低碳策略研究组、可持续研究组、项目团队与外部可持续发展专业人士的共同努力，其研究成果在2020年被纳入"可持续战略手册"。自此，ZHA沿着三个主要方向拓展研究内容：项目参与和评估、知识及工具开发、制订目标及研究报告。碳评估流程表明人类对气候挑战有了更加深入的认知，且正在为降低项目运营能源成本、使用不同结构解决方案和探索新材料的应用做出极大努力。

本次展览展示了ZHA用于评估项目全周期可持续性设计的关键指标，以便实现设计的有效调整并将碳排放影响降至最低。因项目的差异，某些指标的相关性和有效性有所不同。

Sustainability

The climate emergency and carbon consumption are global concerns that extend beyond manmade boundaries, necessitating urgent remedies for the impact of the built environment. ZHA's commitment to sustainability reinforces the prime directive of providing shelter and fostering societal development across generations.

ZHA's ambition is to continue to play a leading role in resolving the critical challenge of global warming through constant improvement in the Studio's work. Projects undergo assessments for their carbon use against internationally recognised targets to understand their current impacts, identify best practices, and determine areas requiring drastic changes—ensuring a consistent delivery of better outcomes for each new project.

The Sustainability Team at ZHA applies detailed environmental analysis and geometry optimisation techniques to maximise project efficiencies, while ZHA's Low Carbon Group promotes more resource-efficient designs, reducing waste and carbon impacts. As a company, ZHA advocates knowledge sharing and strategies that reduce operational and embodied carbon within architecture, with the ambition that today's exceptional project becomes tomorrow's new standard.

The collaborative efforts of the in-house Low Carbon Group, Sustainability Team, and project teams, as well as external sustainability experts, were synthesised into a Strategy Handbook in 2020. Since then, ZHA has been working on expanding the collective knowledge along three main strands: Engage and Assess, Develop Knowledge and Tools, Set-up Targets and Reporting. The carbon assessment process shows there is now a greater consciousness of the climate challenge, and significant efforts are presently being made to reduce projects' operational energy costs, use different structural solutions, and explore alternative materials.

In this exhibition, key criteria used within the practice are presented as references of metrics to look at throughout project stages, in order to affect change and minimise carbon impact. Depending on the project, some criteria are more relevant or efficient to work with than others.

ZHA 全球可持续策略
ZHA Global Sustainability Strategies

认证流程与减排措施

创造可持续建筑是一个复杂的过程，需要从整体上考虑多个方面。每一个方面都与其他方面紧密相连，且与建筑设计密切相关。ZHA 从人本角度考虑所有可持续性方面，并致力于设计符合最高标准的优秀建筑。通过这样的做法，ZHA 能确保他们设计的建筑将在减少环境影响和碳足迹的同时，提供更高水平的舒适度。

Certification Process and Emissions Reductions

Creating sustainable architecture is a complex process that requires looking at multiple areas holistically. Each one of them is interconnected with the other, and they are intimately linked to the design of the building. ZHA considers all the aspects of sustainability from a human centric perspective and works on designing the best buildings to meet the highest possible standards. By doing so ZHA ensures that their buildings will have a reduced environmental impact and lower carbon footprint while achieving higher levels of comfort.

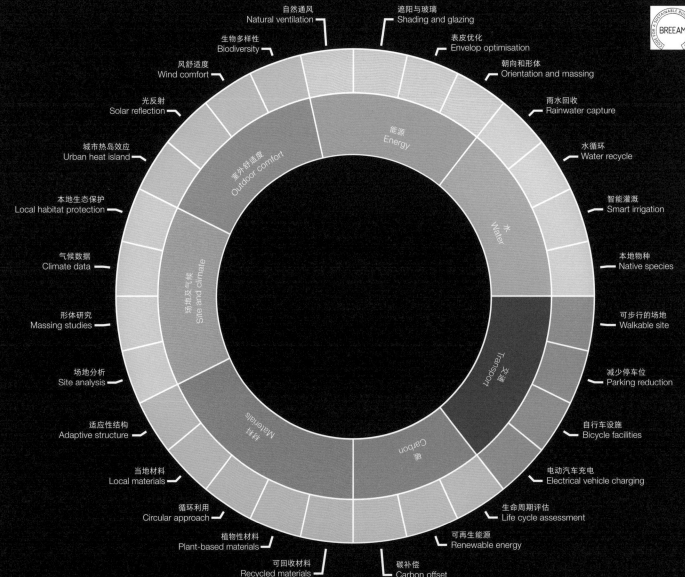

	太阳能研究 Solar studies
	循环经济 Circular economy
	排放 Emissions
	净零能耗 Net zero
	可再生能源 Renewables

运用气候数据建模的弹性建筑
Resilient architecture modelled
with climate data

Net Zero Targets

Internationally known for some of the world's most exciting buildings, ZHA's portfolio of built work is testament to the application of its rigorous design approach. ZHA embraces the complexity and challenges inherent in creating state-of-the-art designs. Using a 'Science-Driven Design' approach, ZHA works with environmental analysis tools to develop the sustainable buildings of the future. In order to reduce carbon footprint and climate change contribution at ZHA, the teams take informed decisions based on the environmental performance of a building to target Net Zero solutions that could improve comfort and minimise environmental impact.

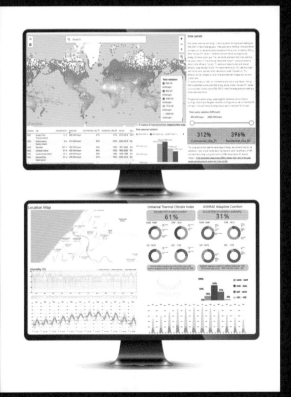

运用气候数据建模的弹性建筑
Resilient architecture modelled with climate data

气候数据分析

ZHA可持续发展团队的工作包括开展研究和开发工具，以追求更具可持续性的设计实践。一个关键的研究领域是利用气候数据进行建筑物理模拟。通过采用这种方法，团队可以评估能否实现不同可持续性认证所概述的目标。通过综合性研究，该团队可以优化设计解决方案，尽可能获得最高的可持续性分数并推广节能建筑类型，同时最大限度地减少碳足迹。通过模拟各种场景并仔细分析得到的数据，他们可以找到提高设计性能和可持续性的机会。

Climate Data Analysis

ZHA sustainability team work involves conducting research and developing tools to support the pursuit of more sustainable design practices. A key research area focuses on the utilisation of climate data to run building physics simulations. By employing this approach, the team can assess whether the objectives outlined by different sustainability certifications will be achieved. Through their comprehensive studies, the team can optimise design solutions to attain the highest possible sustainability scores and promote energy-efficient architectural typologies, all while minimising carbon footprints. By simulating various scenarios and carefully analysing the resulting data, they can identify opportunities to enhance the performance and sustainability of their designs.

事务所项目碳审计
Carbon Assessment of ZHA Projects

ZHA根据国际和区域气候目标衡量他们的项目。实现这些目标意味着要放弃传统的施工技术,并需要改变ZHA的设计和交付方法。ZHA通过使用不同的结构解决方案和替代材料来尽力降低项目的运营能源成本。

Zaha Hadid Architects assess their projects against international and regional climate targets. Reaching these targets means moving on from typical construction techniques and requires a shift in ZHA's design and delivery approach. Significant efforts are made to reduce projects' operational energy costs, to use different structural solutions, and alternative materials.

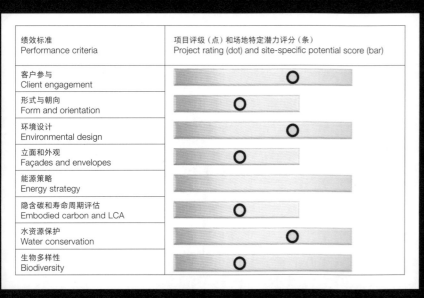

对关键绩效标准进行分析
Key performance criteria are analysed

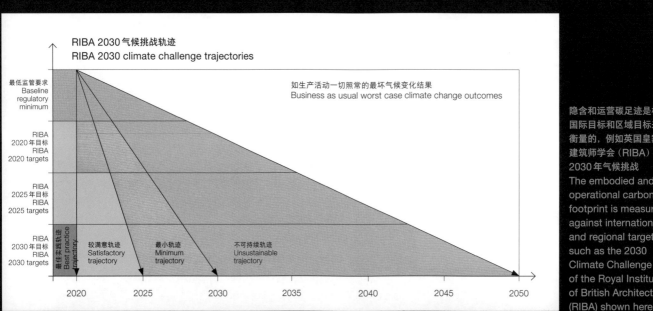

隐含和运营碳足迹是根据国际目标和区域目标来衡量的,例如英国皇家建筑师学会(RIBA)的2030年气候挑战
The embodied and operational carbon footprint is measured against international and regional targets, such as the 2030 Climate Challenge of the Royal Institute of British Architects (RIBA) shown here

项目：伊拉克中央银行
Project: Central Bank of Iraq

位置：伊拉克，巴格达
Location: Baghdad, Iraq

阶段：概念设计 | 方案设计 | 设计深化 | 施工监督
Stage: CD | SD | DD | CS

框架：混凝土/钢
Frame: Concrete/Steel

幕墙：釉面/实体/百叶
Façade: Glazed/Solid/Louvred

运营能源
Operational Energy

千瓦时/平方米/年（处理后的建筑面积）
kWh/m²/year (Treated Floor Area)

100 125 150 175 200 225 250 275 300 325 350 375 400 425 450 475 500 525 550

ECON-19 最优实践
Econ 19 Best Practice

ECON-19

摘录自能源消耗指南19——办公空间能源使用。从碳信托获得
Taken from Energy Consumption Guide 19–Energy Use in Offices. Availavble from The Carbon Trust

上图展示了位于伊拉克的一个办公类项目的运营碳分析评估。根据当地相关指标ECON-19，-525～550是可接受的最大运营碳数值，325～525是业界迄今为止最好的数值。ZHA的伊拉克中央银行所得数值在100～125之间，远低于业界最好的数值。

Example showing the Operational Carbon Analysis Assessment for an office typology case study project in Iraq. According to the relevant local target scale, ECON-19, 525–550 is the largest acceptable Operational Energy value, 325–525 is the industry's best to-date. ZHA's Central Bank of Iraq achieves significantly lower values, between 100–125.

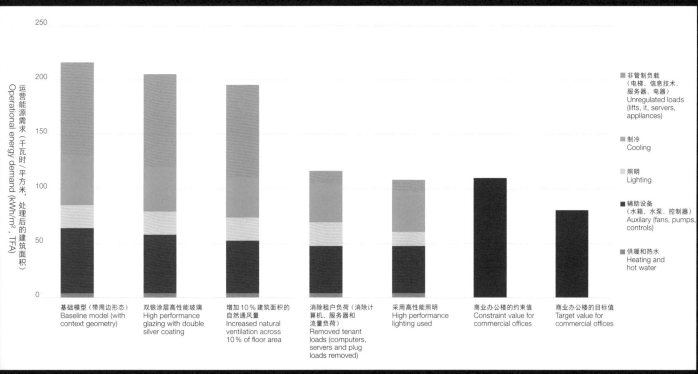

设计优化
通过确定可持续性变化和替代方案，并优化各种提案，建筑设计能够满足所需的限制条件。基础模型是评估时的建筑设计。ZHA可以提出不同的技术方案，以实现较低水平的运营能源需求

Optimising the Design
Sustainable changes and alternatives are identified, and through various proposed optimisations, the building design is able to meet the desired constraints. The baseline model is the building design at the moment that it is assessed. Through different technical options, we can propose to achieve lower levels of operational energy demands

日照研究和遮阳策略

日照研究涉及分析与特定地点相关的太阳位置和太阳辐射模式。在建筑设计的早期阶段进行日照研究至关重要，有助于制定有效的遮阳策略，以减少夏季的过度热量获取，同时允许冬季获取太阳辐射能量。通过日照研究，可以深入了解太阳全年的运动轨迹及其与该地点的相互作用。这些信息有助于确定最佳建筑朝向、形体和开窗设计，以优化能源效率和热舒适性。

Solar Study and Shading Strategies

Solar study involves the analysis of the sun's position and solar radiation patterns in relation to a specific site. The integration of solar study in the early design stages is crucial. It allows for the development of effective shading strategies that mitigate excessive heat gain in summer while allowing for solar heat gain in winter. Insights into the sun's movement throughout the year and its interaction with the site are gained through conducting a solar study. This information enables the determination of optimal building orientations, massing, and windows design to optimise energy efficiency and thermal comfort.

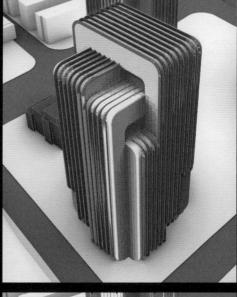

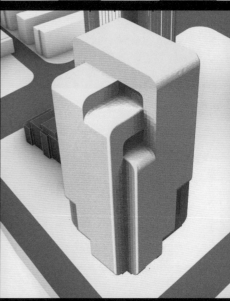

中节能·上海首座项目，中国，上海
太阳辐射
CECEP Shanghai, Shanghai, China
Solar Radiation

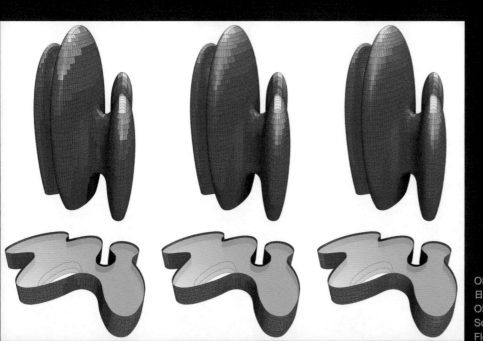

OPPO 国际总部，中国，深圳
日照研究—上图：塔楼；下图：100米高处的楼层细节
OPPO Headquarters, Shenzhen, China
Solar Studies - Top: View of the Tower; Bottom: Floor Detail at 100m Height

遮阳

在了解当地环境条件的前提下,为受太阳直射的部分玻璃幕墙设计遮阳系统,并优先考虑外部遮阳。有选择地使用遮阳系统可降低夏季的高日照辐射热量,同时在冬季实现被动采暖。采用适应性遮阳元素,以便根据采光及遮阳需求及时调整,从而保持平衡。

Shading

The aim is to provide solar shading on all glazed façades receiving direct sun during occupied hours, with a preference for external shading. Selective use of shading can eliminate high solar gain in summer while allowing passive heating in winter. Adaptable elements are very efficient. The need for shading must be balanced against access to daylight and wellbeing.

塔希提国际会议中心和剧院,法属波利尼西亚,塔希提岛
雨篷在提供遮阳的同时,也是每个场地门厅的延伸。它们在整个场地和景观中创造了一个连续的长廊,并结合当地茂盛的植被提供遮阳和雨水回收系统,起到自然通风和降温的作用

International Convention Centre and Theatres, Tahiti Island, French Polynesia
Canopies serve as both shade providers and extensions of the foyers for each venue. Their succession creates a continuous shaded promenade throughout the entire site and landscape. Complemented by the natural shading from the vibrant local vegetation and the implementation of a recycled rainwater system, these canopies actively contribute to the natural ventilation and cooling of the overall design

MAXXI：国立二十一世纪美术馆，
意大利，罗马
优化的屋顶形态确保画廊拥有良好
的采光，同时避免阳光的直射

MAXXI: Museum of XXI
Century Art, Rome, Italy
The optimised roof form
ensures ample daylight for
the galleries while effectively
shielding them from direct
sunlight

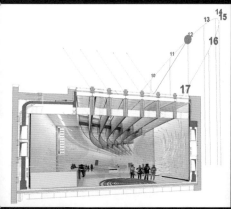

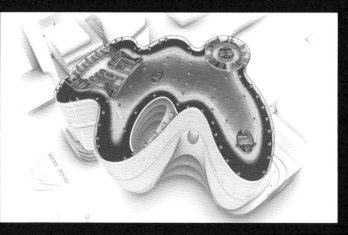

OPPO 国际总部，中国，深圳
日照模拟
OPPO Headquarters, Shenzhen, China
Daylight Simulation

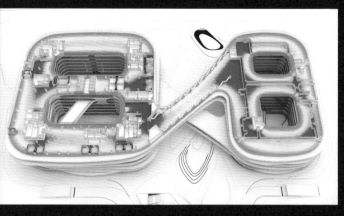

广州无限极广场，中国，广州
日照模拟
Guangzhou Infinitus Plaza, Guangzhou, China
Daylight Simulation

光分析

光分析是 ZHA 的一个重要研究领域，重点是了解和优化建筑物中的自然采光。ZHA 认识到采光对能源消耗和居住者的巨大好处，因此将其列为可持续设计的重要组成部分。有效利用自然光有助于减少白天室内空间对人工照明的依赖，进而减少能源消耗。通过利用先进的模拟技术和设计策略，可以优化建筑布局、窗户位置和遮阳设备。这些措施旨在最大限度地提高自然光透过率，同时最大限度地减少如眩光和得热等带来的诸多问题，从而为居住者带来更健康、更优质的室内环境。

Daylight Studies

Daylight analysis is a key area of research at ZHA, with a focus on understanding and optimising natural lighting in buildings. The team acknowledges the substantial benefits of daylighting for energy consumption and occupant wellbeing, making it an essential component of sustainable design. Efficient use of daylight assists in the reduction of energy consumption by minimising reliance on artificial lighting during daylight hours. Through the utilisation of advanced simulation techniques and design strategies, building

阿卜杜拉国王石油研究中心，
沙特阿拉伯，利雅得
一系列的遮阳庭院穿插在整个建筑中，
从而巧妙控制光照
King Abdullah Petroleum Studies

朝向
了解当地环境条件，分析各种方案，确定最有利的方向，
以减少太阳辐射和立面的风压，提供遮阴和避风雨的地方。

Orientation

Understand local environmental conditions, analyse
options to identify the most favourable orientation to
reduce solar gain, wind pressures on façades, and to
provide shade and shelter from wind and rain.

伊拉克中央银行，伊拉克，巴格达
测试和优化塔楼的朝向，打造了节能、自遮阳的围护结构
Central Bank of Iraq, Baghdad, Iraq
Testing and optimising the orientation of the tower led to an
energy efficient, self-shaded envelope

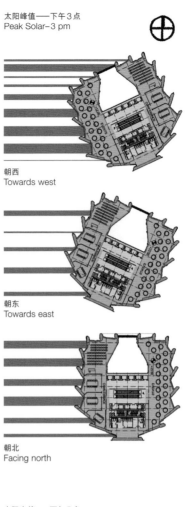

太阳峰值——下午3点
Peak Solar–3 pm

朝西
Towards west

朝东
Towards east

朝北
Facing north

太阳中值——下午5点
Median Solar–5 pm

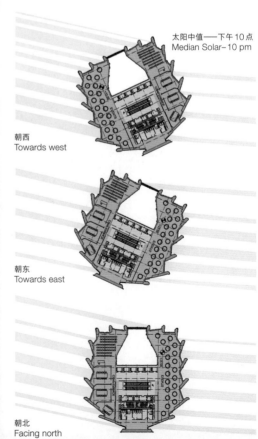

太阳中值——下午10点
Median Solar–10 pm

朝西
Towards west

朝东
Towards east

朝北
Facing north

泾河新城国际文化艺术中心，
中国，西安
风环境模拟
Jinghe New City Culture and
Art Centre, Xi'an, China
Wind Simulation

广州无限极广场，中国，广州
空气对流
Guangzhou Infinitus Plaza,
Guangzhou, China
Cross Ventilation

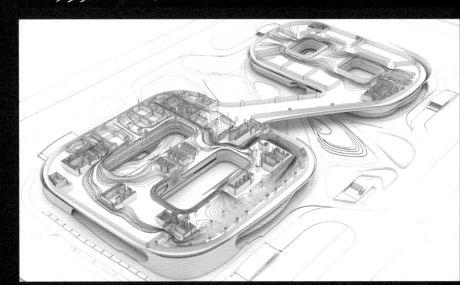

风环境舒适度和自然通风

计算流体动力学（CFD）是一种用于分析建筑物内部和周围流体流动行为及其相互作用的数值技术。CFD是ZHA研究中不可或缺的一部分，它提供了一种了解和预测风如何在建筑物和结构周围流动的方法。通过模拟风的模式，可以评估和优化设计的性能，同时兼顾可持续性和居住者的舒适度。通过检查建筑物周围的风型，可以确定因风速过大而导致的潜在的不舒适区域。这些知识可以帮助我们优化建筑形体，并结合防风林或空气动力学元素等特征，最终提高居住者的舒适度。

Wind Comfort and Natural Ventilation

Computational Fluid Dynamics (CFD) is a numerical technique used to analyse the behaviour of fluid flows and their interaction inside and around buildings. CFD serves as an integral part of ZHA's research, providing a means to understand and predict how wind flows around buildings and structures. Through the simulation of wind patterns, the performance of designs can be evaluated and optimised with considerations for sustainability and occupant comfort. By examining wind patterns surrounding buildings, potential areas of discomfort resulting from excessive wind speeds can be identified. This knowledge enables the optimisation of building shape and the incorporation of features like windbreaks or aerodynamic elements, ultimately leading to improved comfort for occupants.

牛津大学圣安东尼学院中东中心，
英国，牛津
尽可能采用自然通风，特别是通过
礼堂座椅下方的热交换空间
Investcorp Building for Oxford
University at St Antony's
College, Oxford, UK
Natural ventilation is used
wherever possible and
particularly through a labyrinth
below the auditorium seating

自然通风

有效的自然通风策略旨在确保用户控制的通风口（如窗户）和自动通风系统（由 CO_2 或温度传感器触发）达到最高效率。在低层和高层都提供开口，可以最大限度地利用浮力效应，并利用横流通风的风压。此外，建议采用混合模式（自然通风和机械通风）系统以满足全年的设计条件并减轻客户对潜在风险的担忧。

Natural Ventilation

An effective natural ventilation strategy aims to secure user-controlled vent openings (such as windows) and automatic systems triggered by CO_2 or temprature sensors) for maximum efficiency. By providing openings at both low and high levels, the strategy maximises the benefit of buoyancy effects and utilises wind pressures through cross-flow ventilation. Additionally, a mixed-mode (natural and mechanical ventilation) system is proposed to meet design conditions throughout the year and alleviate any concerns about perceived risks for the client.

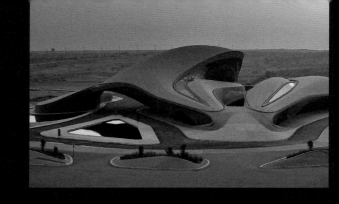

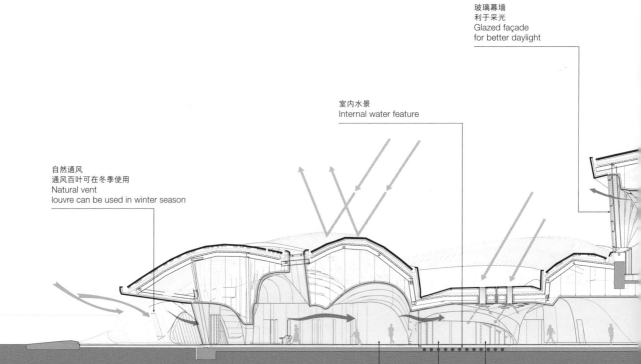

自然通风
通风百叶可在冬季使用
Natural vent
louvre can be used in winter season

室内水景
Internal water feature

玻璃幕墙
利于采光
Glazed façade
for better daylight

循环热缓冲区
降低室内制冷需求
Circulation
thermal buffer zones reduce internal area cooling requirement

室外庭院遮阳系统在冬季及过渡季为员工和访客提供舒适的户外环境体
Well-shaded courtyard
provides reasonable outdoor environme
staff and visitors in winter and mid-seas

在庭院中为员工和访客提供室外水景
External water features in courtyards for staff and vis

碧哈总部，阿联酋，沙迦
BEEAH Headquarters, Sharjah, UAE

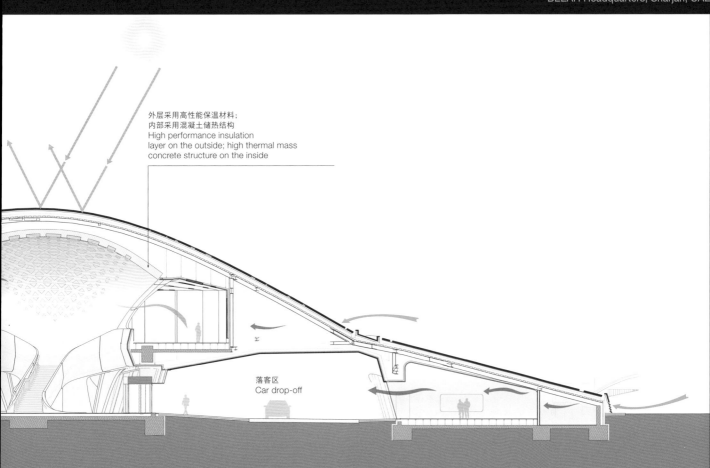

外层采用高性能保温材料；
内部采用混凝土储热结构
High performance insulation layer on the outside; high thermal mass concrete structure on the inside

落客区
Car drop-off

involves considering factors such as sunpath, building design, and materials. This allows to understand the timing and locations where concentrated reflections are likely to arise, thus assessing the extent and impact of solar reflection on surrounding spaces. Based on the results, the team develops strategies to address solar reflection and glare. This involves integrating shading devices like louvres or fins to redirect or diffuse sunlight. Another option is to enhance the façade surfaces by using less reflective materials or applying coatings to reduce solar reflection. Additionally, strategic placement of trees, canopies, or other physical barriers is considered to block direct sunlight and minimise glare.

热点研究
Hot spots studies

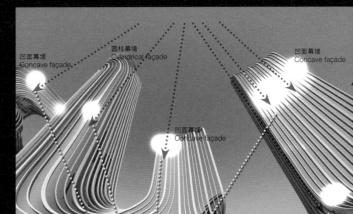

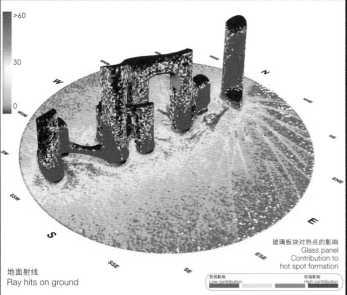

地面射线
Ray hits on ground

玻璃板块对热点的影响
Glass panel Contribution to hot spot formation

低弱影响　较强影响
Low contribution　High contribution

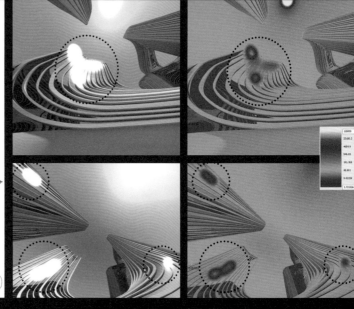

热点研究
Hot spots studies

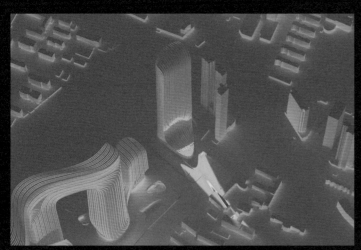

减少热点
Hot spots mitigation

幕墙类型

幕墙应响应内部空间的需求。立面设计决定了自然采光和通风的机会。开口和窗户的位置应尽早纳入设计中。考虑材料、饰面和施工技术，以减少材料用量和浪费，同时提高热性能。

Façade Typologies

Façades should respond to the needs of the internal spaces. Opportunities for daylight and natural ventilation are dictated by façade design. Location of openings and windows should be included early in the design. Consider materiality, finishes, and construction techniques to reduce quantities, waste and improves thermal performance.

达飞海运集团总部，法国，马赛
双层玻璃幕墙与主动冷梁的结合，在降低了运营能耗的同时减少了 23 % 的太阳能得热
CMA CGM Headquarters, Marseille, France
A double skin façade combined with active chilled beams allows 23 % reduction of solar gain while reducing the operational energy consumption

忠利集团大厦，意大利，米兰
外墙由带有隔热板边缘的全高双层外墙组成。双层幕墙中的空腔满足自然通风条件，同时为安装高效的遮阳百叶提供可能，在保护设备免受风雨影响的同时，显著减少了建筑的太阳能得热
Generali Tower, Milan, Italy
The façade is made up of a full height double skin façade with insulated slab edge. The cavity between the façade is naturally ventilated. The double skin façades allows for efficient external deployable blinds, significantly reducing solar gain in the building while devices are protected from wind and rain

形态性能

高效的形式和清晰的细节是关键。双层玻璃的热损失约为实体单元的七倍，因此应仔细考虑玻璃朝向和范围的关系。设计时应注意当地施工技术的局限性，这会对建筑性能产生重大影响。

Envelope Performance

Efficient form and clear detailing are key. Double glazing has around seven times the heat loss of solid elements so the extent of the glazing in relation to orientation should be carefully considered. Design should be mindful of the limitations of local construction techniques which can have a big effect on performance.

格拉斯哥交通博物馆，英国，格拉斯哥
幕墙节点设计和规格严格满足气密性要求
Glasgow Riverside Museum of Transport, Glasgow, UK
The façade detailing and specifications comply with very strict air leakage rates

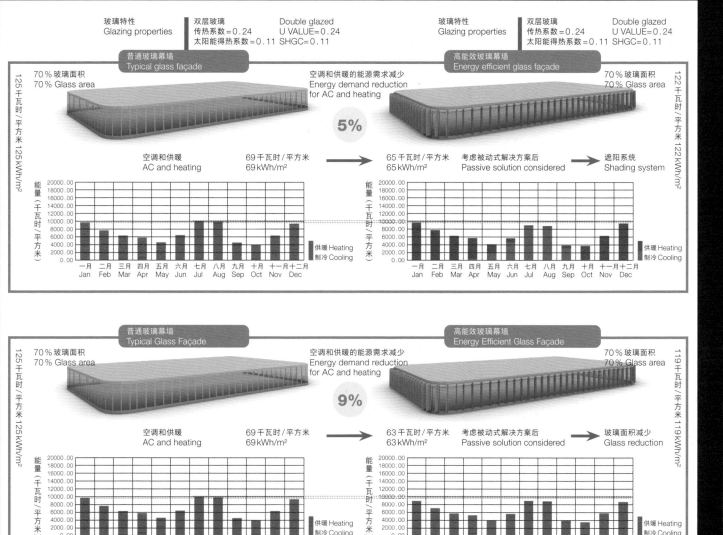
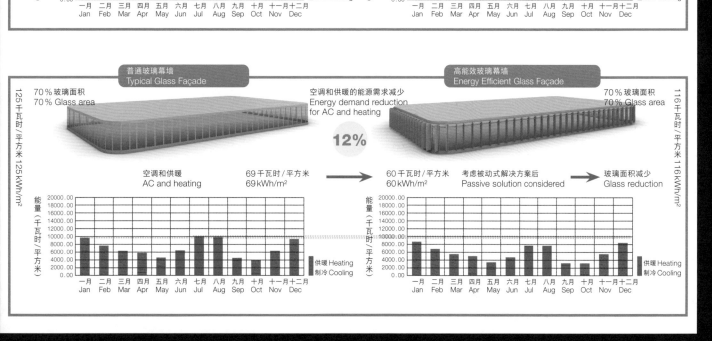

节能策略建模

丽泽SOHO,中国,北京
高效的结构体系服务于建筑高度。
两个体块相互作用,从而减少结构
所需材料

Leeza SOHO, Beijing, China
Highly efficient structural frame for the height of the building. The two core structures are able to support each other, resulting in less material required for the structural frame

塔楼连接和立柱分布
Tower Connection and Column Layout

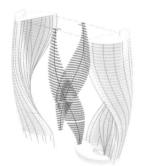

塔楼连接:推力屋顶连接和展廊
Tower Connection: Thrust roof connection and galleries

底部分布
Bottom Layout

顶部分布
Top Layout

结构原理
Structural Principle

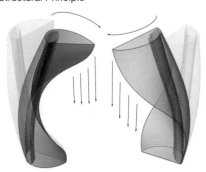

就结构本身而言,每个塔楼都会由于悬臂及其不对称的形状而扭转和弯曲。只有使它们结合为一体,相互借力,才能稳定屹立
On its own, each of the two towers would twist and bend due to its cantilevers and asymmetric shape. Only together they can stand up, when learning against each other

结构连接处保持两个塔楼的平衡
Balance of the two towers by structural connections

在顶层剪力连接,中庭幕墙结构与其铰接
Shear connection at top level, pin connection around atrium façade

结构效率

一级和二级结构的碳含量占建筑物碳含量的比例高达75%。优化风荷载可以将高层建筑结构减少20%。高效的跨度、较小的悬臂和优化的结构板都能减少碳含量。骨骼和纤维等仿生形式非常有效。充气模板、3D打印可用于减少结构中过多的混凝土和金属。

Structural Efficiency

Primary and secondary structures make up as much as 75% of the embodied carbon of a building. Optimising for wind loading can reduce the structure of tall buildings by 20%. Efficient spans, smaller cantilevers, and optimised structural slabs reduce embodied carbon. Bio-mimetic forms such as bones and fibres are very efficient. Inflatable formwork, 3D printing can be used to reduce excessive concrete and metal in the structure.

热质量

石头、砖块、大理石或混凝土等耐热材料有助于缓解温度的波动。即使纳入建筑物的送风路径，饰面仍可发挥有效作用。热质量策略适合搭配自然通风策略，尤其是当建筑物可以在夜间通风（预冷）时。地下迷宫可用于预冷和预热从外部进入的空气。在本页的图片中，有一些利用热质量特性的建筑物示例，例如，阿联酋的碧哈总部、丹麦的奥德鲁普加德博物馆扩建项目和德国的费诺科学中心。

Thermal Mass

Materials with high thermal mass capacities such as stone, brick, marble or concrete help absorb swings in temperatures. Finishes can still be effective if included in the supply air path for the building. Thermal mass works well with natural ventilation strategies, especially when the building can be vented at night (pre-cooling). Underground labyrinths can be used to pre-cool and pre-heat incoming air. In the photographs, there are some examples of buildings that made use of thermal mass properties like, BEEAH Headquarters in UAE, Ordrupgaard Museum Extension project in Denmark, and Phaeno Science Centre in Germany.

碧哈总部，阿联酋，沙迦
BEEAH Headquarters, Sharjah, UAE

奥德鲁普加德博物馆扩建项目，丹麦，哥本哈根
在该项目中，墙壁和天花为裸露的混凝土，以便创造稳定的内部使用环境，并减少运营阶段碳排放
Ordrupgaard Museum Extension, Copenhagen, Denmark
In this project, concrete is exposed in the walls and soffit, stabilising the internal environment and reducing operational carbon emissions

费诺科学中心，德国，沃尔夫斯堡
Phaeno Science Centre, Wolfsburg, Germany

照明和机电策略
建筑设计应以被动调节照明为出发点，以减少对基础设施和能源消耗方面的需求。仔细确定设备点位及管线布局，从而实现工作路径和工作量最小化，并加入高效节能设计（如按需通风、夜间供冷、节能照明、混合模式通风）。

Lighting and ME Strategy
Building design should begin by passively conditioning spaces to reduce the need for infrastructure and energy consumption, carefully locating plant spaces and connections to minimise service runs and sizes, and considering energy efficient services (eg. demand driven ventilation, night cooling, energy efficient lighting, mixed mode ventilation).

碧哈总部，阿联酋 沙迦
BEEAH Headquarters, Sharjah, UAE

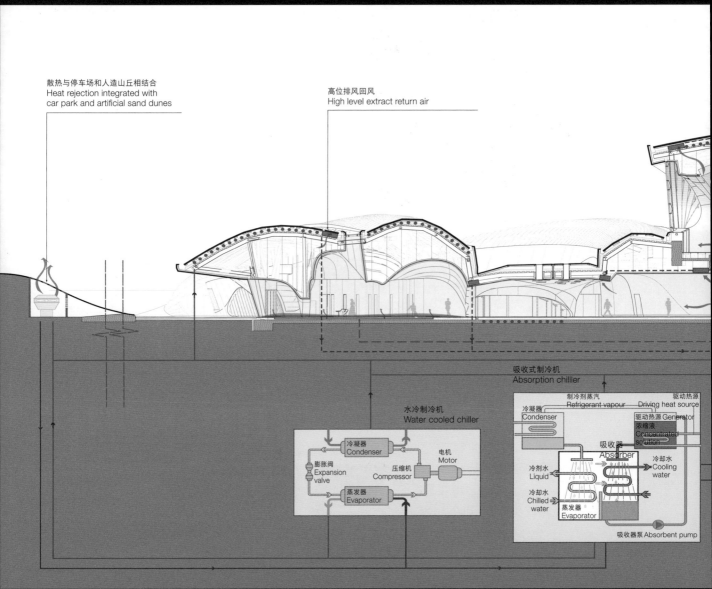

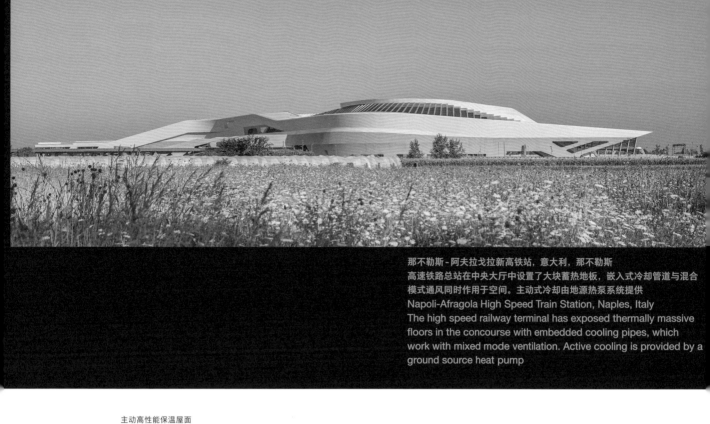

那不勒斯-阿夫拉戈拉新高铁站,意大利,那不勒斯
高速铁路总站在中央大厅中设置了大块蓄热地板,嵌入式冷却管道与混合模式通风同时作用于空间。主动式冷却由地源热泵系统提供
Napoli-Afragola High Speed Train Station, Naples, Italy
The high speed railway terminal has exposed thermally massive floors in the concourse with embedded cooling pipes, which work with mixed mode ventilation. Active cooling is provided by a ground source heat pump

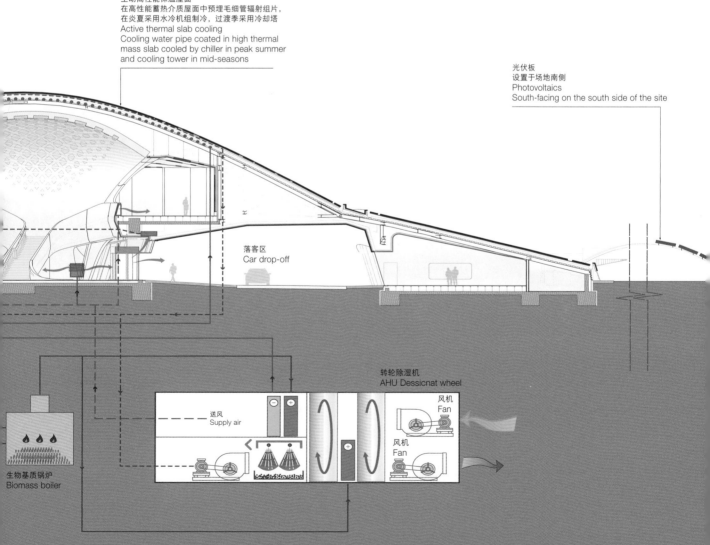

主动高性能保温屋面
在高性能蓄热介质屋面中预埋毛细管辐射组片,在炎夏采用水冷机组制冷,过渡季采用冷却塔
Active thermal slab cooling
Cooling water pipe coated in high thermal mass slab cooled by chiller in peak summer and cooling tower in mid-seasons

光伏板
设置于场地南侧
Photovoltaics
South-facing on the south side of the site

落客区
Car drop-off

转轮除湿机
AHU Dessicnat wheel

送风
Supply air

风机
Fan

风机
Fan

生物基质锅炉
Biomass boiler

低碳材料研究

材料研究是可持续性研究的一个重要方面，重点是要确定适合每个特定项目地点的当地可持续材料。该方法包括通过早期生命周期评估（LCA）考虑隐含碳和全球变暖等潜在因素来比较材料。使用LCA可以评估材料的整个生命周期，包括提取、生产、使用和报废处理。通过比较不同的材料，可以做出明智的决定，最大限度地减少隐含碳和使全球变暖的可能性。结合循环经济原则，设计过程强调尽可能重复使用现有材料、减少废物，并兼顾适应性和灵活性。优先考虑模块化建筑元素和结构，便于拆卸和未来再利用，以减少资源消耗和产生废物。

Low Carbon Material Research

Material research is a critical aspect of sustainability research, with a focus on identifying local and sustainable materials suitable for each specific project location. The approach involves considering factors such as embodied carbon and global warming potential through early-stage Life Cycle Assessment (LCA) to compare materials. The use of LCA enables the assessment of materials' entire life cycle, including extraction, production, use, and end-of-life disposal. By comparing different materials, informed decisions can be made to minimise embodied carbon and global warming potential. Incorporating circular economy principles, the design process emphasises the reuse of existing materials whenever possible, waste reduction, and the integration of adaptability and flexibility. Prioritising modular building elements and structures facilitates easy disassembly and future reuse, reducing resource consumption and waste generation.

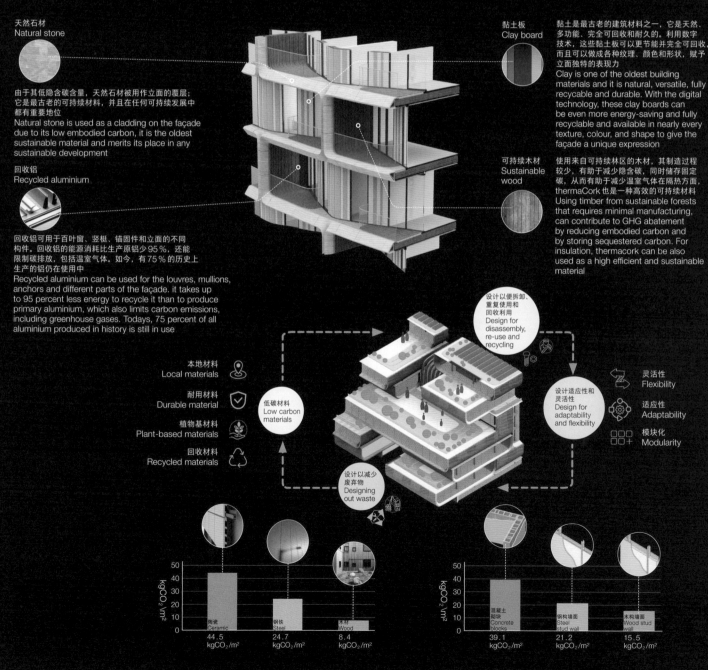

天然石材 / Natural stone

由于其低隐含碳含量，天然石材被用作立面的覆层；它是最古老的可持续材料，并且在任何可持续发展中都有重要地位
Natural stone is used as a cladding on the façade due to its low embodied carbon, it is the oldest sustainable material and merits its place in any sustainable development

回收铝 / Recycled aluminium

回收铝可用于百叶窗、竖梃、锚固件和立面的不同构件。回收铝的能源消耗比生产原铝少95%，还能限制碳排放，包括温室气体。如今，有75%的历史上生产的铝仍在使用中
Recycled aluminium can be used for the louvres, mullions, anchors and different parts of the façade. it takes up to 95 percent less energy to recycle it than to produce primary aluminium, which also limits carbon emissions, including greenhouse gases. Todays, 75 percent of all aluminium produced in history is still in use

黏土板 / Clay board

黏土是最古老的建筑材料之一，它是天然、多功能、完全可回收和耐久的。利用数字技术，这些黏土板可以更节能并完全可回收，而且可以做成各种纹理、颜色和形状，赋予立面独特的表现力
Clay is one of the oldest building materials and it is natural, versatile, fully recyclable and durable. With the digital technology, these clay boards can be even more energy-saving and fully recyclable and available in nearly every texture, colour, and shape to give the façade a unique expression

可持续木材 / Sustainable wood

使用来自可持续林区的木材，其制造过程较少，有助于减少隐含碳，同时储存固定碳，从而有助于减少温室气体在隔热方面，thermaCork 也是一种高效的可持续材料
Using timber from sustainable forests that requires minimal manufacturing, can contribute to GHG abatement by reducing embodied carbon and by storing sequestered carbon. For insulation, thermacork can be also used as a high efficient and sustainable material

本地材料 / Local materials
耐用材料 / Durable material
植物基材料 / Plant-based materials
回收材料 / Recycled materials

低碳材料 / Low carbon materials

设计以便拆卸、重复使用和回收利用 / Design for disassembly, re-use and recycling

设计适应性和灵活性 / Design for adaptability and flexibility

设计以减少废弃物 / Designing out waste

灵活性 / Flexibility
适应性 / Adaptability
模块化 / Modularity

陶瓷 Ceramic — 44.5 kgCO₂/m²
钢铁 Steel — 24.7 kgCO₂/m²
木材 Wood — 8.4 kgCO₂/m²

混凝土砌块 Concrete blocks — 39.1 kgCO₂/m²
钢构墙面 Steel stud wall — 21.2 kgCO₂/m²
木构墙面 Wood stud wall — 15.5 kgCO₂/m²

伦敦水上运动中心，英国，伦敦
大部分外露的混凝土混合了40%的煤副产品——粒化高炉矿渣粉（GGBS）
London Aquatics Centre, London, UK
Most of the exposed concrete incorporated 40 % ground granulated blast slag (GGBS), a by-product from coal

材料循环利用和改造

在可能的情况下，ZHA会为项目指定当地的材料。材料和产品的运输通常占"从起始地到现场"碳排放量的5%。有机和天然材料通常具有较低的隐含碳。大部分的固有能量体现在结构系统中，所以在现有框架上重新覆层是可持续的。

Materials Circularity and Retrofit

Where possible ZHA specifies local materials for our projects. Transport of materials and products typically contributes 5% of the Cradle to Site carbon emissions. Organic and natural materials generally have lower embodied carbon. Most of the embodied energy is in the structural system, so recladding over an existing frame can be sustainable.

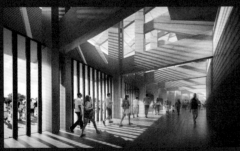

沃克斯豪尔十字岛，英国，伦敦
该立面被设计为零部件组合，采用模块化设计，以简化和减少废弃自遮蔽的外包结构
Vauxhall Cross Island, London, UK
The façade was designed as kit of parts with a modular design in order to simplify and reduce waste self-shaded envelope

安特卫普港屋顶大楼，比利时，安特卫普
比利时的安特卫普港屋顶大楼的地上总面积达12,800平方米，其中一半的建筑由一个历史保护消防站建筑翻新而成
Port House, Antwep, Belgium
Porthouse in Antwerp, Belgium totals 12,800 m² above ground with half of the area being made up by the refurbishment of a redundant fire station

格兰森林流浪者生态足球场，英国，斯特劳德
因地制宜，采用周边森林中的木材，打造木结构建筑。使用轻薄覆层和低能耗材料，让建筑的碳排放远远低于目标水平
Forest Green Rovers Eco Park Stadium, Stroud, UK
Timber Structure, lightweight cladding, and low embodied materials resulted in a design with embodied carbon far surpassing the good practice targets

可再生能源

可再生能源整合旨在减少对不可持续的电力资源和化石燃料的依赖。结合现场生产可再生能源在这一努力中发挥着关键作用,有助于减少碳排放并向低碳未来过渡。太阳能光伏系统用于将阳光转化为电能。通过在屋顶和立面上安装光伏系统,可以利用丰富的太阳能,在现场生产清洁、可再生的电力。此外,在可行的情况下,还可以探索风力发电。在设计中加入风力涡轮机,可以充分利用现场合适的风力条件。

Renewable Energy

Renewable energy integration aims to reduce dependency on unsustainable power resources and fossil fuels. The incorporation of on-site renewable energy production plays a pivotal role in this endeavour, contributing to the reduction of carbon emissions and the transition to a low-carbon future. Solar photovoltaic systems are utilised to convert sunlight into electricity. By incorporating PV systems on rooftops and façades, abundant solar energy is harnessed, enabling the generation of clean and renewable electricity on-site. Furthermore, when viable, exploration of wind power generation is undertaken. Integration of wind turbines into the design capitalises on suitable on-site wind conditions.

风速
Wind speed

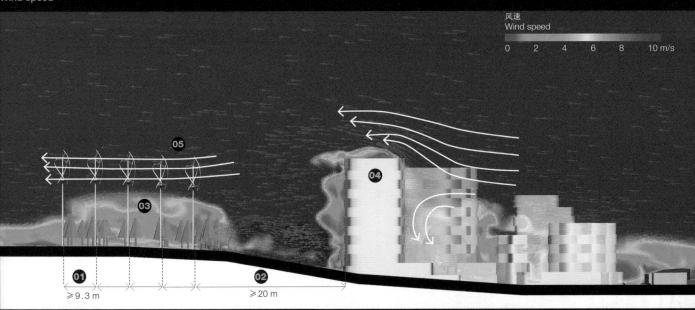

01	风力涡轮机间距	02	噪声	03	树木	04	建筑物	05	风力条件
	涡轮机之间的距离应至少为9.3米,以尽量减少相互干扰		在距离建筑物20米的地方布置涡轮机有助于保持建筑物周围稳定的声环境		有树的区域将会使风速降低,涡轮机应高于树木		当空气在建筑物上方流动时,风速会随之加速,这将增加发电量		涡轮机通常的运行速度在10米/秒~16米/秒,并在20米/秒时切断以保护自己
	Wind turbine spacing Turbines should be located at least 9.3 m apart from each others to minimise interference		**Noise** Allocating turbines 20 m apart from buildings help to keep calm environment around the buildings		**Trees** The area with trees will reduce wind speed at the trees level, turbines should be higher than the trees in these areas		**Buildings** When the air flows above the buildings, the wind speed will be accelerated after. This will boost the generated electricity		**Wind conditions** Turbines usually run in the 10 m/s to 16 m/s, and cuts out to protect itself at 20 m/s

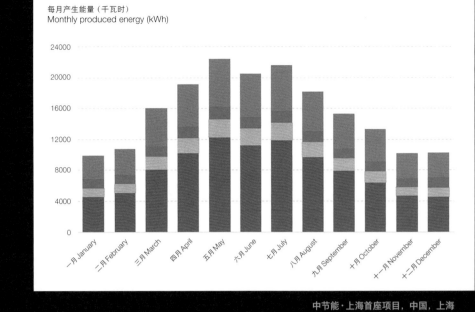

中节能·上海首座项目，中国，上海
各月份光伏能源生产
CECEP Shanghai, Shanghai, China
Monthly Produced Solar Energy

中节能·上海首座项目，中国，上海
光伏板位置及能源生产
CECEP Shanghai, Shanghai, China
Solar Panels Location and Energy Production

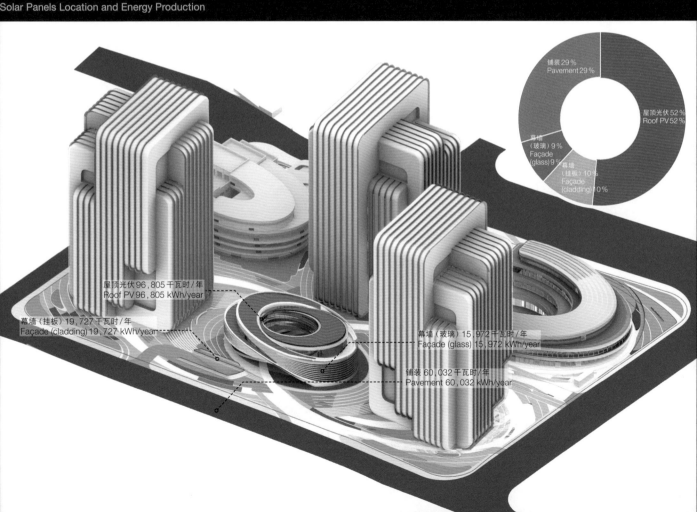

可再生能源

根据场地、气候、位置和公共设施网络的碳排放强度，结合建筑设计寿命内的未来气候情况，评估不同类型可再生能源的可行性和适用性。如具备光伏发电的可能性，应尽量扩大覆盖范围。项目在满足最低规划要求的基础上，应努力提升能源认证水平。

Renewable Energy Sources

Assess the viability and availability of using different renewable energy sources based on site, climate, location, and carbon intensity of utility networks. Consider future climate scenarios over the design life of buildings. If PVs are viable, coverage should be maximised. Projects should endeavour to go beyond minimum planning or certification requirements.

大兴国际机场，中国，北京
机场设计旨在通过在停车场上方设置的光伏板，满足电力总需求的10
Beijing Daxing International Airport, Beijing, China
10 % of the total electrical energy demand of the airport was planned to be met by photovoltaic panels over the carpark.

忠利集团大厦，意大利，米兰
Generali Tower, Milan, Italy

碧哈总部，阿联酋，沙迦
BEEAH Headquarters, Sharjah, UAE

水和生物多样性

每个项目都应在水利用方面做到系统、零排放。冷凝水和处理废水（污水和食物垃圾）也可循环利用，用于灌溉和冲洗厕所。结合景观设计，雨水收集可进一步保证水资源的合理开发、高效利用。

Water and Biodiversity

Projects could aim to be a net zero consumer of clean water. Condensate water as well as processed waste water (black water and food waste) could be recycled for irrigation and WC flushing too. The landscape could be planned to be further water retaining and contribute to cleaning waste water.

中节能·上海首座项目，中国，上海
CECEP Shanghai, Shanghai, China

格林森林自由人生态足球场，英国，斯特劳德
Forest Green Rovers Eco Park Stadium, Stroud, UK

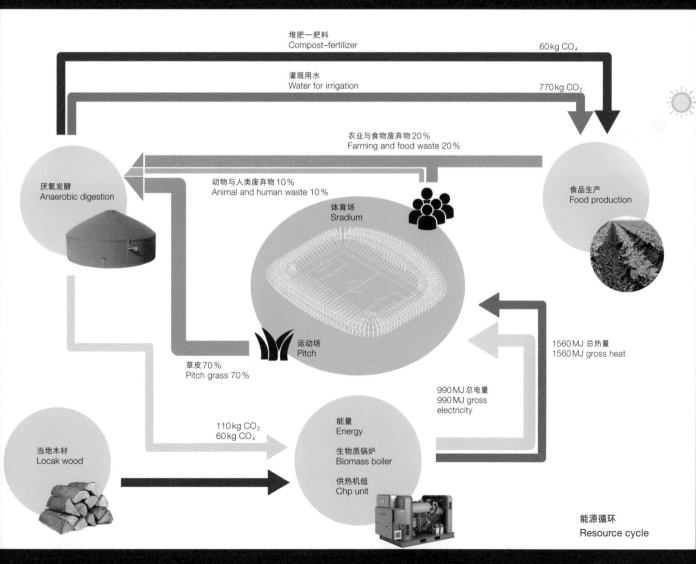

超级天际线

"超级天际线"是由扎哈·哈迪德建筑事务所（ZHA）主导的一个参数研究装置，用以探索高层建筑设计与建造的新概念。该装置展示了一些已建成与未实现的高层建筑项目，显示出ZHA对高层建筑结构的广泛研究和设计。

ZHA的设计为世界各地的城市天际线添砖加瓦。这些设计将水平街道活动平滑过渡到在垂直分层的塔楼结构里进行的活动，无一不强调与周围城市环境的完美融合。

Super Skyline

'Super Skyline,' is an installation featuring the parametric research led by Zaha Hadid Architects, exploring new concepts for designing and constructing high-rise buildings. This installation presents a selection of both realised and unrealised high-rise projects, a testament to ZHA's extensive research and design of high-rise structures.

ZHA's designs inform city skylines across the world. With an emphasis on seamlessly integrating with the surrounding urban context, the designs transition from horizontal street-level activities to vertically layered and activated arrangements within the tower structure.

启程中国

作为扎哈·哈迪德建筑事务所（ZHA）发展的重要里程碑，这次展览庆祝了ZHA在中国的15年旅程。这一切都始于一个非凡的序幕——2010年，ZHA在亚洲的第一座建筑——广州大剧院的竣工。这座享誉全球的建筑被《纽约时报》誉为"同时代世界上最迷人的歌剧院"。这一开创性的成就使ZHA在中国跻身为享有盛誉的建筑公司，开启了ZHA在中国的旅程，并在充满活力的中国建筑领域留下了永恒的作品。

2014年，ZHA设计的香港理工大学设计学院赛马会创新楼正式竣工，为ZHA的亚洲项目又添一项成就。这也成了后续一系列为中国建筑蓝图做出贡献的项目的催化剂。ZHA在北京和上海连续四次与SOHO中国合作，共同开发了超过1,000,000平方米的建筑，与中国的联系不断加强。此后，ZHA通过北京、长沙、南京和其他城市的大型文化、办公、基础设施和酒店项目在中国建筑行业产生了积极反响。

ZHA通过建筑和都市主义的设计手法深入研究中国当代生活模式的复杂性，并采用创新的设计理念和先进的技术手段来组织和处理这些复杂条件。与20世纪建筑中普遍存在的重复性和分离性不同，ZHA的建筑积极参与、整合并适应周围环境。通过开发有助于组织和规划工作、教育、娱乐、居住和交通等多方面日常生活的系统，ZHA创造了一种清晰且具有导向性的生活秩序。

大自然是ZHA重要的灵感来源。ZHA在打造建筑环境时借鉴了自然的系统性、连贯性和美学。虽然ZHA建筑流畅的形式常为公众所惊叹，但其关注的焦点并不局限于美学。ZHA对每个项目的建设目标都会进行深入研究和解读，并不断探索、革新设计方法来优化使用者的体验。

基于与中国各地工程师的合作，结合先进材料和施工技术，ZHA才得以解决关键的生态可持续性问题。建筑行业高度综合的行业特性和知识、技术使与之相关的各个行业共同发展，这也正是ZHA的愿景。近年来，中国坚定不移地致力于发展世界级的文化、民用和交通项目，为人民的未来服务，这种雄心壮志和奉献精神为繁荣城市铺平了道路。

Zaha Hadid Architects in China

Marking a significant milestone in architectural history, ZHA proudly celebrates their 15-year journey in China with this exhibition. It all began with a remarkable prologue—the completion of ZHA's first building in Asia, the Guangzhou Opera House, in 2010. Esteemed globally, this building earned high praise from the *New York Times* as 'the most alluring opera house built anywhere in the world in decades.' This groundbreaking achievement established ZHA alongside renowned architectural firms in China, paving the way for a host of future endeavours and leaving a lasting legacy within the vibrant realm of Chinese architecture.

In 2014, ZHA added another achievement to their repertoire of Asian projects with the completion of the Jockey Club Innovation Tower for the Hong Kong Polytechnic University School of Design. This accomplishment served as the catalyst for further built work that would significantly contribute to the evolving architectural landscape in China. Strengthening the ties between ZHA and China, ZHA received four consecutive commissions encompassing over 1,000,000 square metres across Beijing and Shanghai, developed in collaboration with SOHO China. The firm's impact subsequently reverberated through major cultural, office, infrastructure, and hospitality schemes in Beijing, Changsha, Nanjing, and other cities across China.

ZHA embodies an approach to architecture and urbanism that delves into the complexities of contemporary life patterns in China, employing innovative design concepts and new digital methods to organise and navigate these intricacies. Departing from the repetition and separation prevalent in previous century's structures, ZHA's buildings actively engage, integrate, and adapt to their surroundings. By developing systems that facilitate the organisation and planning of multifaceted life processes encompassing work, education, entertainment, habitation, and transportation, ZHA creates an order that maintains legibility and orientation.

Nature serves as a significant source of inspiration, as ZHA draws upon its systems, coherence, and beauty when crafting built environments. While the public often appreciates the stunning fluid forms of ZHA's architecture, the focus extends beyond mere aesthetics. Each project involves a thorough investigation and interpretation of an institution's purpose, with a constant exploration of new and improved ways to enhance the experience of those who utilise the spaces.

Collaboration with engineers throughout China, combined with the utilisation of advanced materials and construction techniques, empowers ZHA to address crucial ecological sustainability issues. The construction industry's profound knowledge and skills enable remarkable advancements across various sectors, aligning with ZHA's vision. China, in recent years, has demonstrated an unwavering commitment to developing world-class cultural, civic, and transportation projects that will serve its people well into the future. This ambition and dedication pave the way for flourishing cities.

广州歌剧院

中国，广州
2003—2010年
广州市政府

广州大剧院独特的"双砾"设计将相邻的文化建筑与广州珠江新城的国际金融中心相结合，促进了广州的文化发展，增强了城市魅力。大剧院共设1800个座位，配有最新的自然和数字声学技术设备；此外，还有一个400座的多功能厅，可满足表演艺术、歌剧和音乐会等多种演出形式的需要。

设计的灵感来自河谷地貌，尤其是河流侵蚀所带来的地形改变。引申于这种地形的棱线定义了大剧院的室内外区域，在建筑内外切割出引人注目的"峡谷"，用作动线、大堂及咖啡馆，并让自然光渗透到建筑深处。不同元素和不同楼层之间的平滑过渡延续了这一景观设计语言。剧院内部采用定制的玻璃纤维增强石膏（GFRG），延续了流畅、无缝的建筑语言。

Guangzhou Opera House

Guangzhou, China
2003–2010
Guangzhou Municipal Government

The Guangzhou Opera House is a catalyst for Guangzhou's cultural development. Its unique twin-boulder design enhances the city by opening it to the Pearl River and unifying the adjacent cultural buildings with the international finance towers Guangzhou's Zhujiang New Town. The 1800-seat auditorium of the Opera House features the latest acoustic technology, and the 400-seat multifunction hall is designed for performance art, opera and concerts in the round.

The design is inspired by the topography of river valleys, particularly the way they are transformed by erosion. Fold lines in this landscape define territories and zones within the Opera House, cutting dramatic interior and exterior canyons for circulation, lobbies and cafés, and allowing natural light to penetrate deep into the building. Smooth transitions between disparate elements and different levels continue this landscape analogy. Custom moulded glass-fibre reinforced gypsum (GFRG) units have been used for the interior of the auditorium to continue the architectural language of fluidity and seamlessness.

赛马会创新楼

中国，香港
2007—2014 年
香港理工大学

赛马会创新楼（JCIT）是香港理工大学（PolyU）设计学院及以研究为主导的赛马会社会创新设计院的所在地。这座高 15 层、建筑面积 15,000 平方米的大楼可容纳 1800 多名学生和教职员工，并配有以设计教学及创新用途为主的先进设备，包括设计工作室、实验室、工作坊、展览厅、多功能教室、演讲厅和公共休闲区。JCIT 创造了一个新的城市空间，丰富了校园生活的多样性，展现了学校面向未来的无限活力。

大楼位于校园中一个狭窄且不规则的地块上，其设计融合了塔楼和裙楼的特征，形成了更流畅、自由的空间形式。室内外庭院围合出非正式的交流空间，与大型展览论坛、工作室、剧院和娱乐设施相互呼应。设计通过连接设计学院内的各种功能，建立起一种集体研究的文化，使各种学术研究和课题讨论都能够在此相互滋养，从而促进多学科环境的形成。

学生、教职工和访客都可以在这栋 15 层的建筑里自由穿梭。玻璃幕墙和挑空设计增强了内部空间的透明性和连通性，而流通路线和公共区域的布置则鼓励了众多学习组群和设计学科之间的交流与互动。

Jockey Club Innovation Tower

Hong Kong, China
2007–2014
Hong Kong Polytechnic University

The Jockey Club Innovation Tower (JCIT) is home to the Hong Kong Polytechnic University (PolyU) School of Design and the research-led Jockey Club Design Institute for Social Innovation. The 15-storey, 15,000-square metre tower accommodates more than 1,800 students and staff and houses facilities for design education and innovation, including design studios, labs and workshops, exhibition areas, multifunctional classrooms, lecture theatre, and a communal lounge. The JCIT creates a new urban space that enriches the diversity of university life and expresses the dynamism of an institution looking to the future.

Located on a narrow, irregular site of the university campus, the design of the tower dissolves the typical tower and podium building typology into a more fluid composition. Interior and exterior courtyards create informal spaces to meet and interact, complementing the large exhibition forums, studios, theatre, and recreational facilities. The design promotes a multidisciplinary environment by connecting the variety of programs within the School of Design, establishing a collective research culture where various contributions and discourses can nourish each other.

Students, staff, and visitors move through 15 levels of studios, workshops, labs, exhibition, and event areas within the school. Interior glazing and voids bring transparency and connectivity, while circulation routes and communal spaces have been arranged to encourage interaction between the many learning clusters and design disciplines.

凌空SOHO

中国，上海
2010—2014年
SOHO中国有限公司

凌空SOHO位于虹桥商务区内的临空经济园区，毗邻虹桥综合交通枢纽，使凌空SOHO可以便捷地连接地铁和铁路。因此，凌空SOHO在跨国、国内公司与地方企业和社区间的密切联系中发挥着关键作用。

凌空SOHO的设计采用了逻辑清晰的模块化系统，8.4米的结构网格和约18米的建筑进深最大限度地为建筑提供了便利性和灵活性。这座综合性功能建筑的概念围绕着四个平行体块展开，展现了优雅、简约的设计思路。这四个主要元素定义了建筑的外观，蜿蜒的商业裙楼和连续的金属表皮相连接，勾勒出建筑纤长的造型，逐渐向上延伸至屋顶绿化，其建筑形态让人联想到中国书法中力量和舒展的融合。

建筑的四个主要体块向中心汇聚，形成内部庭院，增强了空间体验感。空中连桥则进一步增强了室内外公共空间的动态整合，打造出其独特的魅力。ZHA与SOHO中国在北京、上海共合作过四个获得LEED认证的项目，打造了140万平方米的综合城市空间，凌空SOHO便是其中之一。

Sky SOHO

Shanghai, China
2010–2014
SOHO China Ltd

Situated in the Linkong Economic Park within the Hongqiao Business Zone, Sky SOHO benefits from the exceptional connectivity provided by this transportation hub, with easy access to subway and rail links and close proximity to Shanghai's city centre and the airport. Sky SOHO therefore plays a crucial role in connecting multinational and national enterprises with the local businesses and community.

The design of Sky SOHO provides convenience and flexibility, employing a simple modular system based on an 8.4-metre grid and a building depth of approximately 18 metres. The concept of this mixed-use building revolves around four parallel slabs that embody an essence of elegant simplicity. These four main components define the building's envelope, interconnected by a sinuous retail podium and a continuous metallic ribbon that wraps their elongated form, ultimately culminating in a green roof. The resulting architectural configuration is reminiscent of the blend of strength and fluidity found in Chinese calligraphy.

Strategically arranged, these four primary elements converge at their centre, creating intimate courtyards that enhance the spatial experience. Connecting sky bridges further enhance the integration of dynamic interior and exterior public areas, offering a captivating environment. Sky SOHO is one of ZHA's four collaborations with SOHO China to achieve LEED certification, to talling 1.4 million square metres of mixed-use urban space in Beijing and Shanghai.

南京国际青年文化中心

中国,南京
2011—2018年
南京市河西新城规划部

南京国际青年文化中心建设之初是为了服务于2014年青年奥林匹克运动会,由两座酒店塔楼(其中包括新的南京卓美亚酒店)、一个配备会议设施的文化中心、一个城市广场、办公室和综合功能区组成。该建筑目前处于"青奥会"结束后的运营阶段,继续为提高和振兴城市环境服务,已经成为南京河西新城未来投资的着眼点和催化剂。

文化中心的设计灵感来源于南京底蕴深厚的非物质文化遗产"云锦",这是一种有着1600多年历史的精细提花织造技艺。该中心的设计呈现出一种引人入胜的、形似书法的三维构图,体现着云锦的内涵。一笔连续的线条贯通整个文化中心,与塔楼形体交织连接,并进一步延伸到新中央商务区、滨江公园和江心洲。

两座塔楼分别高达255米和315米,在竣工时是ZHA所设计建成的最高塔楼。其逐渐收窄的轮廓最大限度地扩展了面江的视野,双轿厢电梯则优化了空间利用率并增加了载客量。室内环境的设计旨在提升用户体验,包括充足的自然光、机械与自然通风的有机结合。该项目的建设只用了不到三年,是中国第一座完全采用逆作法建造的塔楼,施工周期缩短了一年。

Nanjing International Youth Cultural Centre

Nanjing, China
2011–2018
Hexi New Town Planning Bureau

Initially developed to accommodate the 2014 Youth Olympic Games, the Nanjing International Youth Cultural Centre comprises two hotel towers, including the new Jumeirah Nanjing Hotel, a cultural centre with conference facilities, an urban plaza, offices and mixed-use areas. Now in its legacy phase, the centre enhances and revitalises its urban environment, serving as an anchor and catalyst for future investment in Nanjing's Hexi New Town.

Taking inspiration from Nanjing's rich heritage of Yunjin, a 1,600-year-old tradition of intricate brocade weaving, the design of the cultural centre manifests as a captivating three-dimensional composition resembling calligraphy. Reflecting the essence of Yunjin, a continuous line interweaves throughout the cultural centre, connecting it with its earthquake-resistant towers and extending further to the new central business district, riverside park, and Jiangxinzhou Island.

The towers, standing at heights of 255 metres and 315 metres, were the tallest completed towers of Zaha Hadid Architects at the time of completion. Their tapering profiles maximize views of the river, while double-deck elevators optimize space utilization and increase passenger capacity. The interior environment is designed to enhance the user experience, incorporating ample natural light, a harmonious blend of mechanical and natural ventilation, and the inclusion of elevated gardens. Constructed in under three years, this project is the first tower in China to be built completely using a bottom-up/top-down construction method, resulting in a one-year reduction in the construction timeline.

长沙梅溪湖
国际文化艺术中心

中国，长沙
2011—2019年
梅溪湖投资（长沙）有限公司

ZHA设计的长沙梅溪湖国际文化艺术中心建筑面积为11.5万平方米，包括MICA当代艺术博物馆、一个容纳1800座席及其配套设施的剧院和一个500座的多功能厅。贯穿整个场地的步行路线与周围的街道相连，组成有机的建筑语言。该中心由三个独立的文化场馆组成，建筑周边的灰空间形成了多个外部庭院，不仅能欣赏到邻近的梅溪湖的景观，还连通了湖上的公园和步道。人行道路在此交会，为户外活动和雕塑展览提供了场地。

长沙历来是一个充满活力的传媒中心，以受欢迎的娱乐和影视制作而闻名，这一精髓在梅溪湖国际文化中心得到充分的体现。作为城市文化地标，该中心延续着长沙颇负盛名的文化遗产，承载着各种各样的艺术活动、表演艺术和影视制作。

为了满足多种演出类型的需求，大剧院提供全面的服务功能及各种附属设施。MICA当代艺术博物馆由八个围绕中庭设置的展览画廊组成，用于大型艺术装置陈设和活动举办，还设有为社区研讨会准备的专属空间，以及演讲厅、咖啡厅和艺术商店等。小剧场（又称多功能厅）可以通过调整舞台机械来满足从小型戏剧表演、时装秀、音乐演出到宴会及商业活动等不同的使用需要。

Changsha Meixihu International Culture and Arts Centre

Changsha, China
2011–2019
Meixihu Investment

The 115,000-square metre Changsha Meixihu International Culture and Arts Centre by Zaha Hadid Architects incorporates a contemporary art museum (MICA), an 1,800-seat theatre with supporting facilities and a 500-seat multipurpose hall. The organic language of the architecture is defined by pedestrian routes that weave through the site to connect with neighbouring streets. Providing views of the adjacent Meixi Lake and giving access to the parks and walking trails on the lake's Festival Island, this ensemble of three separate cultural institutions creates external courtyards, where pedestrian routes intersect for outdoor events and sculpture exhibitions.

Historically known as a vibrant communications hub renowned for its popular performances and television productions, Changsha finds its essence beautifully embodied in the Centre. Upholding the city's esteemed legacy, this cultural landmark serves as a captivating representation, hosting a diverse array of art activities alongside performances and television productions.

Designed to host a wide variety of performing arts, the grand theatre encompasses all necessary front-of-house functions while also housing ancillary facilities. Eight juxtaposed exhibition galleries centre around an atrium for large-scale installations and events, the MICA art museum also includes dedicated spaces for community workshops, a lecture theatre, café and museum shop. The Small Theatre, also known as the multipurpose hall, can be transformed to different configurations to accommodate diverse functions and performances, from small plays, fashion shows and music performances to banquets and commercial events.

对话北京

2008年，扎哈·哈迪德建筑事务所（ZHA）在北京设立了办公室，与这个城市建立了持久的联系，并不断发展。在过去的15年中，ZHA北京办公室已经发展到70名员工，为这座城市贡献了多座独特的地标建筑。

"对话北京"是对北京五个ZHA项目的专注探索，也反映了ZHA对它们与城市本身关系的思考。该展区的平面布局对应北京的环路和中轴线，根据五个项目在城市中的地理位置定位项目模型。银河SOHO占据北京的中心地段，位于东二环沿线；望京SOHO位于望京地区的商业中心；丽泽的北京SOHO上位于丽泽商务区的核心区位；大兴国际机场锚定于历史悠久的中轴线，成为连接北京与世界的"新门户"；新国际展览中心二期位于北京顺义区国际空港新城的核心位置，将迎接当地居民和来自中国及世界各地的游客。这五个项目不断地给城市肌理注入令人惊叹的能量，碰撞出一场关于可能性的全新对话。

Conversation with Beijing

In 2008, ZHA established a representative office in Beijing, forging an enduring connection with the city that continues to unfold. Over the past 15 years, the Beijing Office has grown to 70 employees, contributing to ZHA's distinctive landmark projects within Beijing.

'Conversation with Beijing' is a dedicated exploration of five ZHA projects in Beijing and a reflection on their relationship with the city itself. The floor layout of this exhibition zone mirrors the diagram of Beijing's ring roads and central axis, positioning the project models according to their precise physical locations within the city. Occupying significant locations across the city, Galaxy SOHO is situated along the East Second Ring Road; Wangjing SOHO anchors the commercial centre of the Wangjing district; Leeza SOHO is located in the heart of the Lize Business District; the Beijing Daxing International Airport anchors the historic central axis, standing as the 'new gateway' connecting Beijing to the world; and the New China International Exhibition Centre Phase II, at the core of the International Airport New City in Beijing's Shunyi District, will welcome local residents and visitors from across China and around the world. Continually infusing the urban fabric of the city with awe-inspiring energy, these five projects spark conversations of new possibilities.

银河 SOHO

中国，北京
2008—2012年
SOHO 中国有限公司

位于北京市中心地段的银河SOHO，是ZHA与SOHO中国首次合作的作品。该项目获得过许多杰出奖项，已经成为北京市最受欢迎的地标性建筑之一。银河SOHO的设计灵感来自北京城的宏伟规模，建筑面积为33万平方米，是一个集办公、商业与休闲于一身的综合体，也是这座城市不可或缺的一部分。

银河SOHO由四个连续的、具有流动感的体量组成，它们通过延展的连桥相互分开、融合或连接。这种独特的设计使四个体量在各个方向上相互适应，构建了一个全景式的建筑，没有角落空间或者生硬的转折打破其形式的流动性。项目的内部庭院是中国传统建筑的再现，它们相互连接，在四个体量之间创造了一个由连续开放的空间组成的内部世界。

建筑各层之间相互作用，产生了一种沉浸感与空间包围感。随着走入建筑物内部，人们会发现这些空间都遵循着建筑连续的、曲线的设计主旨。地下一层至地上三层是商业与休闲公共空间，再往上是创新型企业办公空间，顶层则是能欣赏到城市绝妙风景的酒吧、餐馆与咖啡馆等的商业场所。这些独特的功能分区通过内部空间相互连接，且与城市有着密切的联系，使银河SOHO成为北京著名的地标建筑。

Galaxy SOHO

Beijing, China
2008–2012
SOHO China Ltd

As the first of five collaborations between ZHA and SOHO China, the Galaxy SOHO project in central Beijing has received a number of outstanding awards and has become one of the city's most popular landmarks. Inspired by the grand scale of Beijing, Galaxy SOHO is a 330,000m^2 office, retail and entertainment complex that is an integral part of the living city.

The composition of Galaxy SOHO consists of four continuous, flowing volumes that are set apart, fused or linked by stretched bridges. This unique design allows the volumes to adapt to each other in all directions, generating a panoramic architecture free of corners or abrupt transitions that break the fluidity of its form. The interior courts of the project are a reflection of traditional Chinese architecture, where courtyards create an internal world of continuous open spaces. Unlike a composition of rigid blocks, the architecture is instead comprised of volumes that coalesce to create a world of continuous mutual adaptation and fluid movement between each building.

The various levels of the building interact with each other to generate a feeling of being immersed and surrounded by the space. As users enter deeper into the building, they discover more intricate and private spaces that follow the same continuous and curvilinear language as the rest of the building. The three lower floors house public spaces for shopping and entertainment, followed by several floors of workspaces for innovative businesses, and the top floor is dedicated to bars, restaurants and cafés with stunning views of the city. These distinct functions are interconnected through interior spaces that share an intimate connection with the city, establishing Galaxy SOHO as a prominent landmark for Beijing.

望京 SOHO

中国，北京
2009—2014年
SOHO中国有限公司

望京距北京市中心约12公里，最初作为一个住宅区开发于20世纪90年代中期，如今已经转变为一个以创意和IT产业为重点的活跃社区。在过去的20年里，随着北京经济和城市化的显著发展，望京也成为首都东北部地区商业、住宅、交通、教育和文化中心。望京SOHO位于望京的中心，总建筑面积521,265平方米，是一个综合体开发项目，由三座分别为118米、127米和200米高的锥形塔楼组成。这些塔楼被设计成三座交织的"山峰"，将建筑与景观完美融合，其中包括一个60,000平方米的绿地公园，为周边社区提供服务。

望京SOHO的建筑设计响应了城市流线，自然光从各个方向渗透进每座塔楼中。塔楼的流线形态在不同视角下持续变化——在某些视角下表现为独立的建筑，而在其他角度看则是相互连接的整体。随着周边住宅的开发以及与其他各区交通方式的发展，塔楼在走势和布局上引导游客和上班族去往该地块周边的各个交通枢纽。项目充满凝聚力的设计语言为不断发展的望京社区提供了一个标志性的城市定位。

望京SOHO对周边地区来说无疑是锦上添花，既满足了对多样化和灵活办公空间日益增长的需求，又考虑到其对环境的影响，通过提供各类公共设施和受欢迎的休闲空间来提高居民的幸福感。望京SOHO在设计、建设和运营管理中采用了3D建筑信息模型（BIM），降低了能源消耗和排放。该项目获得了美国绿色建筑委员会颁发的LEED金级认证。

Wangjing SOHO

Beijing, China
2009–2014
SOHO China Ltd

Wangjing, originally developed in the mid-1990s as a residential area located approximately 12 kilometres from the city centre, has transformed into a vibrant community with a focus on creative and IT industries. Over the past 20 years, as Beijing underwent significant economic and urban growth, Wangjing emerged as a crucial hub for employment, housing, transportation, education, and culture in the northeastern part of the city. Located at the heart of Wangjing, Wangjing SOHO is a mixed-use development of 521,265 square metres, consisting of three tapering towers 118 metres, 127 metres and 200 metres in height. These towers are designed as three interweaving 'mountains' that seamlessly integrate the building with the landscape, including a new 60,000-square metre public park that serves the surrounding community.

The architectural design of Wangjing SOHO responds to the city's flows and allows natural daylight to permeate each building from all directions. The juxtaposition of the tower's fluid forms continuously changes when viewed from different angles, presenting as individual buildings in some perspectives and as a connected ensemble in others. As the development of housing and transportation links to other parts of the city progresses, the orientation and composition of the towers guide visitors and staff towards the various transport options surrounding the site. The cohesive design of the project serves as an anchor and identity for the growing Wangjing community.

Wangjing SOHO is a valuable addition to the local community, meeting the increasing demand for diverse and flexible office spaces while enhancing the well-being of residents through the provision of public amenities and popular leisure spaces, all while considering its environmental impacts. Employing 3D Building Information Modelling (BIM) in its design, construction, and operational management to reduce energy consumption and emissions, Wangjing SOHO has been awarded LEED Gold certification by the US Green Building Council.

大兴国际机场

中国，北京大兴
2014—2019年
北京新机场建设指挥部

大兴国际机场是北京市的一座新机场，位于市中心以南46公里处，乘坐城市快轨仅20分钟即可到达，十分便利。为满足快速增长的国际出行需求，大兴国际机场应运而生。随着中国交通网络的不断发展，它将成为一处重要交通枢纽，且已为未来的扩建计划预留足够的空间，到2025年旅客吞吐量预计可达到7200万人次。机场客运航站楼占地70万平方米，其中包括一个8万平方米的地面交通中心，直接连接北京市中心、地方铁路服务和国家高速铁路网络，从而带动京津冀地区的经济发展。

　　航站楼紧凑、高效的设计借鉴了中国传统建筑以主庭院为中心的原则，将乘客从出发区、到达区和中转区无缝地引导到航站楼中心大厅。航站楼的拱形屋顶由六根高耸的造型柱支撑，不仅能引入自然光，而且能将乘客引向中心大厅。航站楼紧凑的径向布局最大限度地减少了值机柜台和登机口之间的距离，极大地提高了乘客的使用便利性和操作灵活性，在满足航站楼停放最多数量飞机的同时，确保乘客到达最远登机口的步行时间少于8分钟。

Beijing Daxing International Airport

Daxing, Beijing, China
2014–2019
Beijing New Airport
Construction Headquarters

Beijing Daxing International Airport is a new airport in the Daxing district, located 46km south from the city centre and a convenient 20-minute journey by express train. Developed to meet the rapidly growing demand for international travel, Beijing Daxing Airport will be a major transport hub within the country's growing transport network, accommodating 72 million travellers by 2025 and with future expansion plans already in place. The airport's 700,000-square metre passenger terminal includes an 80,000-square metre ground transportation centre that offers direct connections to Beijing, the national high-speed rail network, and local train services, acting as a catalyst for economic development in Tianjin and Hebei Province.

　　Drawing on the principles of traditional Chinese architecture that centre around a main courtyard, the terminal's compact and efficient design seamlessly navigates passengers through departure, arrival, and transfer zones towards the grand courtyard at the heart of the terminal. The terminal's vaulted roof features six soaring forms that support the structure and brings in natural light, directing all passengers towards the central courtyard. The compact radial layout of the terminal maximises passenger convenience and operational flexibility by minimising distances between check-in and gates, allowing direct parking of the maximum number of aircraft at the terminal, and ensuring a walking time of less than 8 minutes to the farthest boarding gate.

丽泽SOHO

中国，北京
2013—2019年
SOHO中国有限公司

丽泽SOHO坐落于北京西南部的丽泽路，其所在的丽泽商务区作为日益发展的金融和交通枢纽，将市中心与最近建成开放的大兴国际机场连接起来。这座45层高、建筑面积为172,800平方米的塔楼满足了创新型企业对灵活、高效的甲级办公环境的需求。

毗邻城市航站楼综合交通枢纽，丽泽SOHO所在的地块被地铁隧道沿对角线一分为二。横跨这条隧道，塔楼在保持外观连续、完整的情况下被一分为二。两个体块之间的空间上下贯通整幢塔楼，并由此形成了目前世界上最高的通高中庭（高达194.15米）。随着塔楼的攀升，中庭以双螺旋形式逐渐扭转上升，使高层办公区得以俯瞰北面的丽泽路，获得更好的视野。

中庭连接着塔楼内部的空间，其雕塑般扭转的姿态，也有助于引入更多的自然光和多样化的城市景观，形成一个充满活力的建筑空间。这里与北京的交通网络无缝集成，为使用者提供良好的可达性。中庭同时也起到了"热交换"作用，为塔楼内部环境提供良好的通风条件，以及高效的空气净化和过滤作用。

Leeza SOHO

Beijing, China
2013–2019
SOHO China Ltd

Located on Lize Road in southwest Beijing, the Leeza SOHO tower anchors the new Fengtai business district—a growing financial and transport hub between the city centre and the recently opened Beijing Daxing International Airport. This 45-storey 172,800-square metre tower responds to demands from small and medium-sized businesses for flexible and efficient Grade A office space.

Adjacent to the business district's rail station, the site of Leeza SOHO is diagonally dissected by an underground subway service tunnel. Straddling this tunnel, the design of the tower divides its volume into two halves, enclosed by a single façade shell. The emerging space between these two halves extends the full height of the tower, creating the world's tallest atrium at 194.15 metres, which rotates through the building as the tower rises to realign the upper floors with Lize road to the north.

The atrium in Leeza SOHO connects all spaces within the tower, bringing in natural light and diverse views due to its sculptural and twisting form. It acts as a vibrant civic space that seamlessly integrates with Beijing's transport network and provides great accessibility. Serving as a thermal chimney, the atrium ensures excellent ventilation that keeps air clean and filtrates unwanted airflow.

新国际展览中心二期

中国，北京
2021年（建设中）
北京辰星国际会展有限公司

新国际展览中心毗邻北京首都国际机场，现已成为举办会议、交易会及行业博览会的重要场所，吸引着世界各地的代表。为了满足不断增长的空间需求，ZHA设计了新国际展览中心二期工程，该扩建项目占地438,500平方米，将大大增加展览空间，巩固北京作为国际信息交流中心的地位。该中心位于顺义区国际空港新城的核心地带，未来将承办各类综合活动，吸引当地居民以及来自中国和世界各地的游客。

项目的核心设计理念旨在体现展览馆、会议中心和酒店之间的和谐关系。这些功能空间由一系列相互连接的线条和几何图形组成，灵感来源于中国传统建筑中釉面琉璃瓦的纹理；古铜色的主色调与巨型凹槽开窗进一步增强了建筑外表皮的视觉动态效果与表现力。南北中轴作为展厅之间的主要连接，设有一系列功能清晰、灵活可变且实用性强的空间，包括景观花园、咖啡馆以及室外活动空间等，供人们放松休闲或进行非正式会谈。庭院上方的连廊进一步增强了各个空间之间的联系。会议中心和酒店位于场地的北部，行人、货物及车辆三条流线相互独立。这种设计保证了流线的通畅且可避免干扰正在进行中的活动，使场地适应各种使用情况。

复合屋顶系统可为展览中心提供良好的吸音效果。屋顶的对称几何形态形成一系列高效的轻型大跨度结构，打造出室内无柱的灵活空间，可以轻松地满足各种展览需求。屋顶的工业风选材和规模与流动的建筑语言相平衡。模块化的制造与施工技术，使得施工时间、投资和运营成本最小化。

New International Exhibition Centre Phase II

Beijing, China
2021 (Under Construction)
Beijing Chenxing
International Exhibition

Located adjacent to Beijing's Capital International Airport, the New International Exhibition Centre has become a prominent venue for conferences, trade fairs, and industry expos, attracting delegates from around the world. To meet this growing demand, ZHA has designed Phase II, a new expansion spanning 438,500 square metres, which will significantly increase the exhibition space and reinforce the city's status as a premier hub for international knowledge exchange. Situated in the heart of the International Airport New City of Shunyi District, the centre will host a wide range of events, attracting both local residents and visitors from across China and beyond.

The design of the centre reflects a harmonious connection between the exhibition halls, conference centre, and hotel. Drawing inspiration from the textures of glazed tubular ceramic tile roofs in traditional Chinese architecture, the design features a series of interconnecting lines and geometries. The visually dynamic envelope is enhanced by its copper colour and large recessed windows, adding to the expression of the aesthetic. The central north-south axis serves as the primary connecting space in the centre, creating shared courtyards in landscaped gardens, cafés, and outdoor event spaces for informal meetings and relaxation. This spatial layout ensures functional clarity, flexibility, and efficiency. Secondary bridges at higher levels enhance connectivity within the facility. With the conference centre and hotel positioned to the north, the movement of people, goods, and vehicles within the centre is well-organised into three separate routes to facilitate circulation and adaptability, and minimise disruption to ongoing events.

The centre's design incorporates a composite roof system that provides insulation and excellent sound absorption. Its symmetric geometries contribute to a lightweight structure with large spans, offering a flexible, column-free space that can easily accommodate various exhibition requirements. The industrial materiality and scale of the roof are balanced with a fluid architectural language. Modular fabrication and construction methods are employed to minimise construction time, investment, and operational costs.

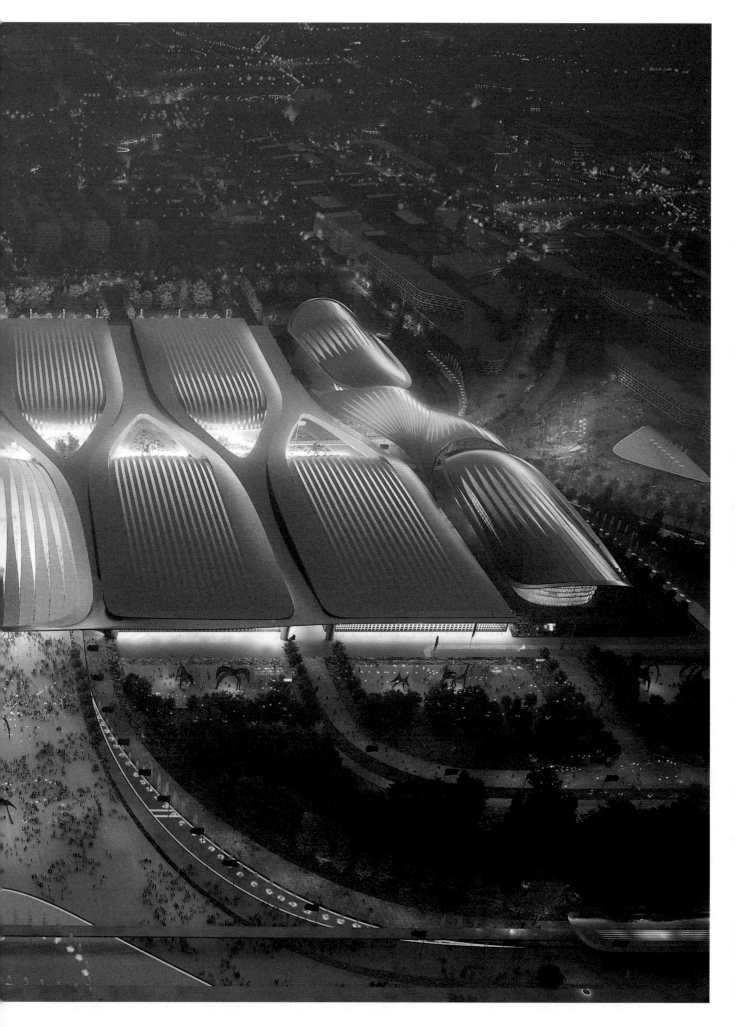

建筑未来

"建筑未来"体现了扎哈·哈迪德建筑事务所（ZHA）对未来城市的深刻见解，展示了中节能·上海首座、深圳湾超级总部基地C塔、深圳OPPO国际总部和成都独角兽岛城市设计等中国正在进行的项目。这些项目遍布中国各地，包括上海、深圳、广州、成都、杭州、西安、武汉、珠海、三亚、香港和澳门等重要城市。目前，ZHA在中国有近30个在建项目，涉及从室内设计到建筑和城市设计等多种类型和体量的项目。

　　本单元展出的项目积极参与着中国的快速发展进程，反映了ZHA对当地文化的理解和对场地特定条件的回应。通过将ZHA在先进技术、材料和建造工艺方面的创新研究与当地环境相结合，这些项目在体现可持续原则的同时成为当地发展的催化剂。在未来的10年里，ZHA的作品将以极高的辨识度和独特的设计语言成为地域文化结构中不可或缺的一部分，持续丰富各地的城市风貌。

　　15年前，ZHA正式进入中国，开启了融合文化传承与未来愿景的创造之旅。ZHA的设计灵感来自中国如画的自然景观和丰富的艺术传统，这些灵感被融入到了事务所的设计之中。ZHA的作品融合了中国传统绘画与园林景观的永恒之美，以及建筑与自然的共生关系，同时展现着对中国的传统、当下和未来的融合。

Building the Future

⑨

Embodying ZHA's insightful perspective on future cities, 'Building the Future' showcases a selection of forthcoming projects in China, such as CECEP Shanghai Campus, Tower C at Shenzhen Bay Super Headquarters Base, OPPO Headquarters, and the Unicorn Island Masterplan. These ongoing projects span in cities all across China, including Shanghai, Shenzhen, Guangzhou, Chengdu, Hangzhou, Xi'an, Wuhan, Zhuhai, Sanya, Hong Kong and Macau to name a few. With an extensive portfolio of nearly thirty ongoing projects throughout various cities, ZHA explores a diverse range of typologies and works across all scales, from interiors to architecture and urban design.

Actively participating in China's rapid development, the exhibited projects reflect an understanding of local culture and a sensitivity to site-specific conditions. By adapting ZHA's innovative research in advanced technology, materials, and construction techniques to the local context, these projects serve as catalysts for local development while embodying sustainable principles. In the forthcoming decade, these projects will become an integral part of the regional cultural fabric, continually enriching the urban landscapes across diverse locations in China.

Reflecting on the significance of their groundbreaking venture into China 15 years ago, ZHA's journey encapsulates the seamless integration of cultural heritage and futuristic aspirations. Their inspiration, drawn from the diverse natural landscapes and rich artistic traditions of China, is woven into their designs. Embracing the timeless beauty of traditional Chinese painting and garden landscapes, together with the symbiotic relationship between architecture and nature, ZHA's work embodies the assimilation of China's past, present, and future.

沐梵世度假酒店

中国，澳门
2013—2018年
新濠博亚娱乐有限公司

位于澳门路氹的沐梵世度假酒店，是综合度假胜地新濠天地的一部分。其设计理念以两座塔楼之间的中庭为核心展开。这一新颖的中庭空间仿佛由一系列透空"雕刻"而成，使得建筑内部能够灵活容纳各种功能并满足客房设施需求。酒店设有770间客房、套房和空中别墅，还设有公共大厅、会议活动设施、游戏室、餐厅、水疗中心、屋顶泳池，以及大规模的后勤服务区域和辅助设施。

受到中国传统玉雕的启发，该设计通过建筑中心的三个镂空结构打造出一扇城市之窗，将内部公共空间与城市景观连接起来。镂空结构定义了酒店的雕塑形态，确保了建筑两侧客房的均匀分布，同时打造出坐拥壮丽城市景观的转角套房。

形态自由的外骨骼结构为建筑提供了结构支撑，因而减少了其对内部支柱的需求，为各种功能分区创造出宽敞、连续的空间。建筑的设计从场地条件和澳门独特的文化遗产出发，融合创新工程，协调形式与功能，为澳门量身打造了一种新的建筑形式。

Morpheus Hotel and Resort at City of Dreams

Macau, China
2013–2018
Melco Resorts and Entertainment

Located in Cotai, Macau, the Morpheus Hotel is part of the leading integrated resort City of Dreams. The design concept centres on a central atrium between two towers, traversed by a series of distinct voids. This unique arrangement allows for flexibility in accommodating a variety of functions and requirements of guest amenities within one building envelope. Housing 770 guest rooms, suites and sky villas, the building also features public lobbies, meeting and event facilities, gaming rooms, restaurants, spa, rooftop pool, as well as extensive back-of-house areas and ancillary facilities.

Inspired by the rich traditions of jade carving in China, the design for the hotel features three external voids carved through the building's centre to create an urban window that connects the interior communal spaces with the city. Defining the sculptural form of the Morpheus Hotel, the voids generate unique corner suites with spectacular views and ensure an equal room distribution on either side of the building.

Morpheus' exoskeleton frame wraps around the exterior, providing structural support that reduces the need for interior columns and creating expansive uninterrupted spaces for diverse functions. Evolved from Macau's particular cultural heritage and site conditions, the hotel's design combines innovative engineering, cohesion form and function, and offers a new architectural language uniquely expressive for the city.

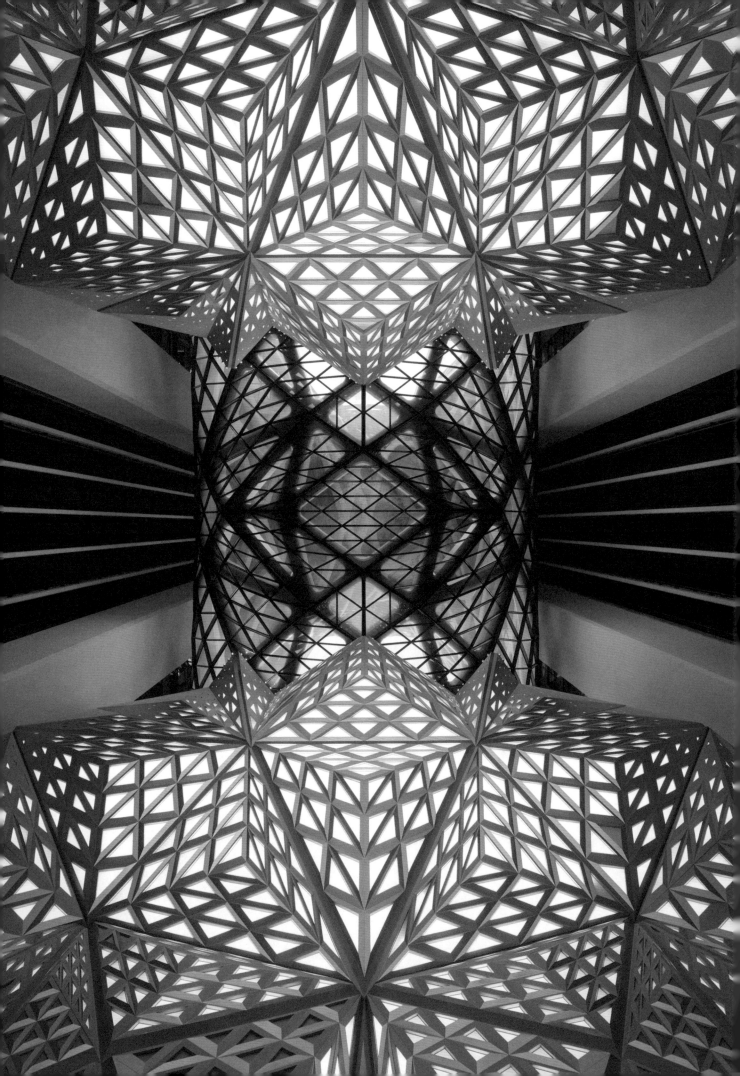

广州无限极广场

中国，广州
2016—2021年
李锦记健康产品集团

广州无限极广场是无限极（中国）新的全球总部，面积达185,643平方米，建在现白云机场原址上，将广州市中心、飞翔公园与机场原址上重建的新社区联系了起来。建筑横跨广州地铁的地下隧道，因此被分为两个体块，通过多层连桥相互连通。

无限极广场共八层，通过一系列"无限"的循环建立协作式工作空间，增强各部门之间的互动和交流。办公空间围绕着室内中庭和室外庭院布局，与无限符号"∞"相呼应，以创造出多层次共享、具有强烈社区感的室内外空间，彰显无限极的企业文化。

在广州湿润的亚热带季风气候下，无限极广场的设计和建造均以获得LEED金级认证和中国绿色建筑三星认证为目标。项目通过结构优化减少了混凝土用量，增加了可回收材料的使用比例——25,088吨的可回收材料被用于建筑施工。

对全年太阳辐射条件的分析帮助ZHA确定了室外露台的进深，以最大化利用建筑的自遮阳形态；同时环境分析也为外立面穿孔铝遮阳板的分布提供数据支持，从而降低太阳能得热、优化外遮阳效果。基于建筑智能管理系统，项目还设置了一个由光伏系统供能的喷雾降温网络，将收集到的雨水雾化后喷洒到中庭上方的ETFE膜屋盖外表面，通过蒸发来实现散热和降温，有效地降低了室内温度并减少能源消耗。

Infinitus Plaza

Guangzhou, China
2016–2021
LKK Health Products Group
(LKKHPG)

Infinitus Plaza is the new global headquarters of Infinitus China. The 185,643-square metre Plaza is built on the site of the decommissioned Baiyun Airport, linking Guangzhou's city centre with Feixiang Park and new communities within the former redevelopment of the airport. Straddling a sub-surface tunnel of the Guangzhou Metro, Infinitus Plaza is divided into two buildings interconnecting at multiple levels.

Establishing collaborative workspaces, Infinitus Plaza is designed over eight storeys as a series of infinite rings that enhance interaction and communication between all departments. Arranged around the central atria and courtyards, the design echoes the infinity symbol '∞,' and creates a variety of shared indoor and outdoor spaces that build a strong sense of community and define Infinitus' corporate culture.

Situated within Guangzhou's subtropical monsoon climate, Infinitus Plaza has been designed and constructed to LEED Gold certification and the equivalent 3-Star rating of China's Green Building Program. Optimisation of the structure has reduced the amount of concrete required and increased the proportion of recyclable content—25,088 tons of recycled materials have been used in its construction.

Annual solar irradiation analysis was used to determine the width of the outdoor terraces that self-shade the building and the deployment of external perforated aluminium shading panels to reduce solar heat gain. Operated by the building's smart management system and powered by photovoltaics, collected rainwater is sprayed onto the translucent, double-layered ETFE membrane roof above each atrium. This process, known as evaporative cooling, effectively lowers interior temperatures and reduces energy requirements.

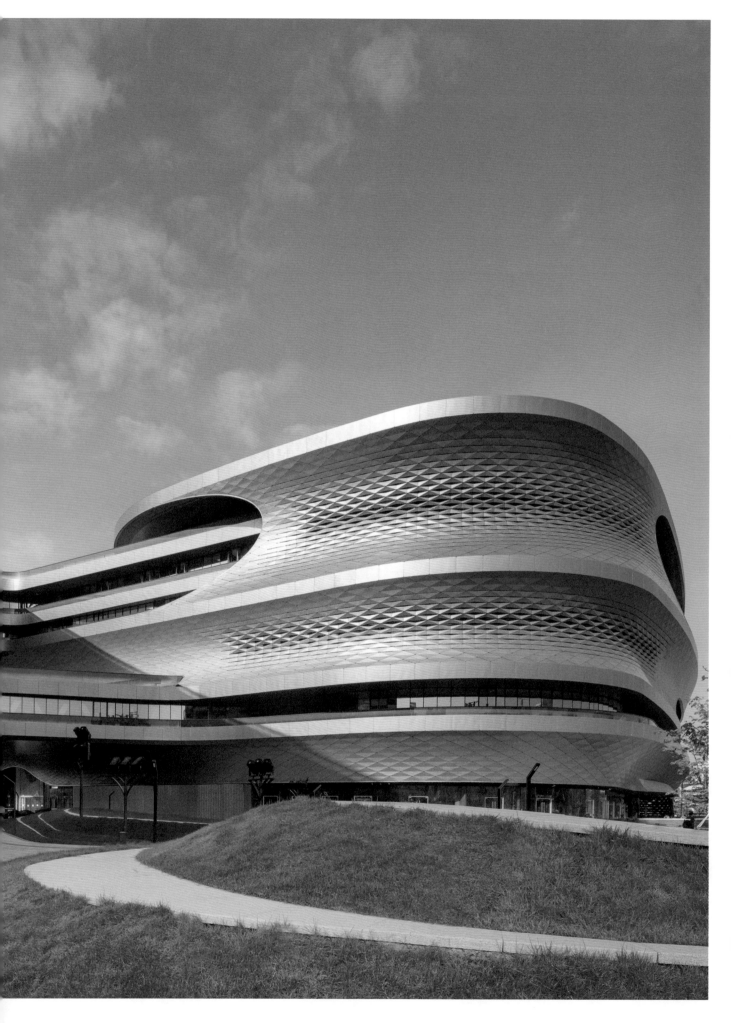

西一线跨绛溪河大桥

中国，成都
2019—2023年
成都国际航空枢纽开发建设有限公司

西一线跨绛溪河大桥是成都天府国际空港新城西段环线的一部分，横跨沱江的主要支流——绛溪河。主跨185米和边跨55米组成了连接两岸总长295米，兼具公路、自行车道和人行天桥功能的西一线跨绛溪河大桥。

西一线跨绛溪河大桥由两个主要钢拱结构组成，该结构可为120~250米跨度的桥梁提供最有效的支撑。拱形结构分别从桥面的两侧升起，向对侧倾斜，并在顶部相接，以稳定桥梁结构，使其免受横向风力的影响。桥墩与桥台富有动感的曲线，与拱形结构和桥面形态完美融合，雕塑般的大桥将成为成都一处地标性的重要交通基础设施。

该桥是用当地制造的预制钢结构现场焊接而成，其设计与施工质量均高于应对200年一遇的极端气象所需的标准。

Chengdu West Line Cross—Jiangxi River Bridge

Chengdu, China
2019–2023
Chengdu
International Aerotropolis Group
Construction
and Development Co., Ltd.

Carrying the western section of the Chengdu Airport New Town ring-road and cycle-route across the Tuojiang River, a leading tributary of Jiangxi River, the 185-metre central span and smaller 55-metre spans connecting to the either riverbank create the 295-metre road, cycle and pedestrian bridge which gently curves as it crosses the river.

As arches provide the most efficient bridge structure for spans between 120—250 metres, Jiangxihe West Bridge is composed of two primary steel arches that rise from either side of the horizontal road deck. The arches lean together to touch tangentially at their crown, stabilising the structure from lateral wind forces. The dynamic curvature of its supporting piers and abutments seamlessly integrates with the primary arches and road deck, defining a sculptural landmark within Chengdu's key transport infrastructure.

Built from locally manufactured prefabricated steel sections spliced together with on-site welding, the bridge's design and construction exceeds the standards required for once-in-200-year weather events.

西安国际足球中心

中国，西安
2020—2023年
陕西西咸中央商务区建设发展有限公司

西安国际足球中心作为一个可以容纳60,000座位的体育场，将用于举办国内与国际赛事、国内联赛、青年训练教学、娱乐演出和文化活动等。体育场以赛后遗产的运营模式为设计的出发点，旨在为足球赛事提供最佳场地，最大限度地增强其功能性。

体育场开放的外立面完美融入沣东商业区的正交城市网格中，可全天提供一系列的公共空间、娱乐与餐饮设施。在足球比赛、文化活动与演出期间，观众可以坐在一系列朝南的遮阳花园露台观赛或观演，同时一览西安城与青山的迷人景色。外立面绵延的横贯线条不仅增强了体育场的美感，还保护它不受北风的侵袭，并突显了屋顶优美的流线型形态。马鞍形的座席则最大限度地增加了中场的观众座位。

足球中心的设计采用超轻的大跨度索网屋顶结构，从而实现了最小的负载和材料消耗，同时减少体育场主要结构的体积。宽阔的屋顶悬于建筑物围护结构内的场地上方，而宽敞的露天平台与公共走廊在各个层面都融入了大量的绿色植物，为西安大陆性气候的炎热夏季创造了舒适的环境。在拉紧的索网结构的支撑下，观众席上方覆盖的半透明膜可保护观众免受恶劣天气和阳光直射的影响。

Xi'an International Football Centre

Xi'an, China
2020–2023
Xixian Central Business District
Construction and Development

The new Xi'an International Football Centre will feature a 60,000-seat stadium designed to host national and international matches, domestic league games, youth training academies, entertainment performances, and cultural events. The design of the stadium prioritises its long-term use beyond the 2023 tournament by providing optimal conditions for football and maximising its functionality.

Integrated within the orthogonal urban grid of Fengdong's business district, the stadium's open façades invite the city into its core, offering a range of public spaces, recreation areas, and dining facilities throughout the day. A series of shaded south-facing garden terraces accommodate spectators during football matches, cultural events and performances while providing stunning views of the city and Qing Mountain. The sweeping lines of the façade not only enhance the stadium's aesthetics but also protect it from northerly winds, while the roof's fluid forms cover a saddle-shaped seating bowl that maximises the spectator seating capacity at midfield.

To minimise the stadium's primary structure, an ultra-lightweight long-span cable-net roof structure is employed in the design. This reduces the load and material footprint while maintaining structural integrity. The generous roof overhangs along the perimeter provide shelter for facilities within the building's envelope, while the spacious open-air terraces and public concourses incorporate extensive greenery at all levels, creating comfortable conditions in Xi'an's hot continental summer climate. Supported by the tensioned cable-net structure, a translucent membrane over the seating protects spectators from inclement weather and direct sunlight.

澳门新濠影汇二期

中国，澳门
2017—2023年
新濠博亚娱乐有限公司

位于澳门路氹的新濠影汇二期，是对现有度假村的扩建项目，在休闲、娱乐和接待方面都为观众提供了无与伦比的体验。受电影元素的启发，此次扩建设计对装饰派艺术时期丰富的细节、醒目的几何图案和精湛的工艺做了现代的诠释。该项目总建筑面积250,000平方米，设有两座酒店塔楼，提供900间客房和套房，拥有21,000平方米的购物和餐饮区域，以及2300平方米的博彩空间。

两座塔楼采用三种不同等级的玻璃和外部装饰元素，有效地减少了太阳热能的吸收和眩光，同时保证了客人的舒适度。高性能的建筑围护结构以及高效的控制系统有助于降低能源需求。此外，设计中使用的所有木材和木制品都通过了林业管理委员会的认证。

考虑到其毗邻澳门路氹生态保护湿地，该项目的构成受环境场地评估的强烈影响。椭圆形塔楼的朝向和布局经过优化，可有效促进度假村内部及周边的自然通风。参与该项目的生态学专家制定了相应措施，以保护该地现有的植物种群，同时补充种植了与景观和栖息地环境相适应的植物种类。新濠影汇二期可持续的建筑设计、开发和管理在2021年英国建筑研究院环境评估方法（BREEAM）大奖中荣获"亚洲区大奖"。

Studio City Phase 2

Macau, China
2017–2023
Melco Resorts and Entertainment

Located in the Cotai district of Macau, Studio City Phase 2 is an expansion of the existing Studio City resort, delivering an unparalleled experience in leisure, entertainment, and hospitality. Inspired by the cinematic references of the resort, the design of this expansion combines the rich detailing, striking geometries, and exquisite craftsmanship of the Art Deco period with a contemporary touch. The 250,000-square metre Studio City Phase 2 features two hotel towers with 900 rooms and suites, 21,000 square metres of shopping and dining areas, and 2,300 square metres of gaming space.

The towers utilise a composition of three glass gradations and external fins, which effectively reduces solar heat gain and glare while ensuring thermal comfort for guests. The high-performance building envelope, efficient services and systems contribute to reducing energy demand. Moreover, all timber and timber-based materials used in the design adhere to Forestry Stewardship Council certification.

Considering its proximity to the protected wetland of the Macau Cotai Ecological Zone, the development's composition is strongly influenced by environmental site assessments. The orientation and configuration of the elliptical towers are optimised to facilitate natural ventilation within and around the resort. The project's ecologist has developed measures to preserve existing flora of the site and carry out compensatory planting of compatible species. The sustainable building design, development, and management of Studio City Phase 2 have been recognised with the 'Regional Award, Asia' at the BREEAM Awards 2021.

香港科技大学学生公寓

中国,香港
2018(建设中)
香港科技大学

香港科技大学是一所领先的研究机构,成立于1991年,一直位居亚洲和全球最佳研究机构之列。香港科技大学也是全球发展最快的大学之一,因此迫切需要在清水湾校区增加新的住宿设施。由ZHA与Leigh & Orange 联合设计的新学生公寓计划于2024年完工,可容纳1500多名学生。该项目将先进的数字设计技术与50年建筑生命周期的可持续实践和运营战略相结合,与香港科技大学利用技术和创新应对当下全球挑战的使命高度契合。

新学生公寓位于校园的东南部,坐落在一处陡峭的坡地上,场地水平高差约25米。建筑屋顶被设计为主要的交通动线,同时结合遮阳措施,为学生及教职工提供休息、交流的场所,这有助于构建更强大的跨校园文化。这条屋顶高线走廊在校区北部的学术区和南部的公寓区之间建立了新的连接,免去了学生和教职工在起伏丘陵地形上的交通烦恼。

建筑设计采用了先进的数字手段,考虑了各种场地参数,如地形高差、太阳辐射、视线和土壤条件,从而优化了建筑布局和朝向。内部空间通过数字编程进行布局模拟测试,以优化空间的功能性和适应性,并准确计算自然光照度。建筑信息模型(BIM)和三维模拟进一步优化了设计协调和材料选择,而与设计深化同步制定的施工策略则确保了高效的采购和施工流程。为了响应香港特区政府提高建筑质量、缩短施工周期和减少建筑垃圾的倡议,这些建筑采用了模块化系统,包括预制幕墙单元和盥洗室模块,便于现场快速安装。

HKUST Student Housing

Hong Kong, China
2018 (Under Construction)
Hong Kong University of
Science and Technology

Established in 1991, Hong Kong University of Science and Technology (HKUST) has emerged as a leading research institution, consistently ranked amongst the best in Asia and worldwide. HKUST is also one of the fastest growing universities globally with an urgent demand to provide new residential facilities within the Clear Water Bay campus. Planned for completion in 2024, the university's new halls of residence designed by Zaha Hadid Architects in collaboration with Leigh and Orange will house over 1,500 students. Marrying advanced digital design technologies with sustainable construction practices and operational strategies for a 50-year life cycle, the design aligns with the university's mission to harness technology and innovation in addressing contemporary global challenges.

Situated in the southeast part of the HKUST campus, the new halls of residence are nestled within a steeply sloping site that spans approximately 25 metres in level difference. The building's roof line has been designed as its primary circulation and incorporates shaded outdoor areas for students and staff to rest and gather, helping to build a stronger cross-campus culture. This elevated rooftop walkway creates a new connection between the academic blocks of the north campus and the primarily residential blocks of the south, eliminating the need for students and staff to circumnavigate the hilly terrain.

The design of the buildings utilises advanced digital design tools that consider various site parameters, such as terrain levels, solar radiation, sightlines, and soil conditions, allowing for an optimised configuration and orientation. The use of digital encoding for internal spaces facilitates layout tests to enhance functionality and adaptability, and enables accurate calculations of natural light levels. Building Information Modelling (BIM) and 3D simulations have further optimised the design coordination and material selection, while construction strategies developed simultaneously with the design development to ensure efficient procurement and construction processes. Aligning with the Hong Kong Government's initiatives for improved build quality and reduced construction time and waste, the buildings' incorporate modular systems, including pre-assembled façade units and washroom pods, facilitating quick on-site installation.

深圳科技馆（新馆）

中国，深圳
2019（建设中）
深圳市建筑工务署

深圳科技馆（新馆）位于光明科学城内，将成为广深科技创新走廊的一大亮点，也是探索科技力量、了解科技对我们生活和未来影响的重要目的地。该馆与中国各地的大学、学校及创新中心联动，将成为新兴的光明科学城至关重要的一部分。受经济政策的鼓舞，深圳科技馆（新馆）将成为具有里程碑意义的机构，也将展示并巩固深圳在创新技术领域的全球领先地位。

深圳科技馆（新馆）占地约125,000平方米，U形的平面布局具有很强的定位感和导向性。围绕着中央庭院，访客可到访参观公共空间、展厅、教育设施等相互连接的空间。精心策划的展览让访客踏上科技馆与城市、自然相结合的探索之旅。以适应性的最大化为基本的设计原则，部分展厅为常设展厅，其他展厅可以根据不同的展览类型灵活改变布局，确保为访客提供多样化和沉浸式的空间体验。

科技馆的设计在体量上对其独特的场地条件做出了回应。建筑以围合的动态曲线几何造型与其东侧的城市交通流线相呼应，并呈现出几乎完全封闭的球体外观。动态的线条沿着科技馆的南北立面逐渐拉长，并在其中"雕刻"出开口，使访客得以瞥见室内丰富的活动景象。建筑向西侧光明公园的自然景观延伸，并在那里渐变为层叠的露台，优雅地勾画出博物馆中央庭院玻璃幕墙的框架。

Shenzhen Science and Technology Museum

Shenzhen, China
2019 (Under Construction)
The Bureau of Public Works
of the Shenzhen Municipality

The Shenzhen Science and Technology Museum, situated within the Guangming Science City, is poised to become a highlight of the Guangzhou—Shenzhen Science Technology Innovation Corridor and a key destination in the region for exploring the power of science and technology and understanding their impact on our lives and future. Linked with universities, schools, and innovation centres across China, the Museum will play an integral role in the emerging Guangming Science City in Shenzhen. Aligning with the larger economic stimulus policy, it will serve as a landmark institution to showcase and strengthen Shenzhen's global position as a frontrunner in innovation and technology.

Spanning approximately 125,000 square metres, the Museum features a U-shaped plan that facilitates intuitive orientation and navigation for visitors. Its interconnected public spaces, galleries, and educational facilities revolve around an atrium courtyard, defining a journey of discovery that harmoniously connects the museum, city, and park through its thoughtfully curated exhibitions. Incorporating maximum adaptability as the basic design principle, certain galleries will maintain a consistent layout, while others will offer flexibility to accommodate different types of exhibitions, ensuring a diverse and immersive spatial experience for visitors.

The design of the Museum presents a volumetric response to the unique site conditions. While effectively conveying the various urban circulation routes towards the east, the building incorporates solid and dynamic curvilinear geometries and takes on the appearance of an almost fully enclosed sphere. Elongating these fluid lines, the museum's north and south side elevations feature carefully carved openings that reveal glimpses into the bustling activity within. The building expands towards the natural landscapes of Guangming Park in the west, in which it creates a series of terraces that elegantly frame the glazed wall leading to the central courtyard atrium at the core of the museum.

成都科幻馆

中国，成都
2022—2023年
成都科创新城投资发展有限公司

成都被山脉和森林所环绕，在其丰富的历史中发展出灿烂的文化。出土的属于3000～5000年前三星堆文明的青铜面具更是激发了人们对神秘幻象和外星文明的想象。今天，成都已发展为一个重要的全球科学创新和研究中心，同时也被公认为中国科幻小说创作的重要孵化地。

成都科幻馆坐落于成都郫都区的菁蓉湖畔，建筑面积为5.9万平方米，设有展览馆、剧院、会议厅及配套的附属设施。屋顶流动的造型从中心向外辐射，犹如一团加速向外扩张的迷人"星云"。沿人行流线设置的一系列活动空间引导访客从城市界面进入场馆，同时将展览、教育设施、咖啡馆及其他设施有序串联。中庭上方的天窗与面朝西岭雪山的巨型玻璃幕墙将自然光线引入建筑，将内部空间与周围环境紧密结合。

2023年，成都科幻馆是第81届世界科幻大会的主会场，这是中国首次举办这一著名活动，具有重要的里程碑意义。该馆还将承办雨果奖颁奖典礼，此奖项以"科幻小说之父"雨果·根斯巴克的名字命名，旨在表彰优秀的科幻小说和奇幻文学作品。这一著名奖项给予了中国科幻小说前所未有的认可，曾获得该奖项的中国作品有刘慈欣的《三体》(2015)和郝景芳的《北京折叠》(2016)。通过举办一系列的活动，成都科幻馆将成为一座科幻文学作品的实体纪念碑，如灯塔般闪耀在艺术创作之巅，为全球科幻社群做出亮眼的贡献。

Chengdu Science Fiction Museum

Chengdu, China
2022–2023
Chengdu Science
and Innovation City Investment
and Development Co.

Surrounded by mountain ranges and forests, the heritage of Chengdu is enriched by the captivating mystique and extraterrestrial motifs discovered in the carvings and masks of the Sanxingdui civilisation, dating back 3,000—5,000 years. Today, Chengdu has transformed into a prominent global hub for scientific innovation and research, while also earning recognition as a thriving incubator for Chinese science fiction.

The 59,000-square metre museum features exhibition galleries, theatre, conference hall, and supporting ancillary spaces. Situated on the Jingrong Lake in Chengdu's Pidu District, the museum floats above the water surface, with its roof featuring fluid forms that radiate from a central point. The design resembles an expanding nebula cloud, with a star at its centre, transforming the museum into a captivating 'star cloud.' Connecting nodes of activity, pedestrian routes guide visitors from the city into the museum, while linking exhibition galleries, educational facilities, cafés, and other amenities. The sky-lit central atrium and the expansive window facing the breathtaking Xiling Mountain connect the museum's interiors with their surrounding environment.

In 2023, the museum takes centre stage as the main venue for the 81th World Science Fiction Convention (Worldcon), marking a significant milestone as China hosts this prestigious event for the first time. The museum will also serve as the esteemed host of the Hugo Awards, named after Hugo Gernsback, sometimes known as 'The Father of Science Fiction.' These renowned awards recognise excellence in science fiction and fantasy literature and have bestowed unprecedented recognition upon Chinese science fiction, with notable recipients such as Liu Cixin's *The Three-Body Problem* (2015) and Hao Jingfang's *Folding Beijing* (2016). The museum's role in hosting these events cements its position as a beacon for the genre and a vital contributor to the global science fiction community.

海南酒店与服务式公寓

中国，三亚
2012年（未建成）
私人

这一总体规划开发项目旨在为海南岛东南海岸建造一个由豪华酒店和服务式公寓组成的综合体。项目坐落于一片私人海滩上，设计灵感来自岛上的自然景观和场地的斜坡地形。建筑综合体遵循土地的自然轮廓，沿着起伏的山坡向下排布。

场地被主路隔开，服务式公寓占据了场地的上半部分；在道路与大海之间的下半部分是一家带有景观花园的五星级酒店。酒店的接待楼位于项目的核心位置，楼内有一个宽敞的、可俯瞰四周的双层大堂。大堂下方有两层专门用于宴会和会议的服务设施，根据需要可提供独立的交通流线。三家特色餐厅与一个泳池酒廊为访客提供海滨餐饮体验以及度假村的全景视野。

酒店的两侧共有235间客房，同时在场地最西端还有22栋别墅。9层的服务式公寓位于场地北侧，公寓东侧有10栋私人别墅，可直接通往海滩与酒店。整体项目通过连桥与通廊无缝连接。

Hainan Hotel and Service Apartment

Sanya, China
2012 (Unbuilt)
Private

Located on the southeast coast of Hainan Island, a southern tropical island in the South China Sea, this masterplanned development unveils an architectural complex consisting of a luxurious hotel and bespoke serviced apartments. Situated on a private beach, the design of the development draws inspiration from the natural landscape of the island and the sloping topography of the site, following the natural contours of the land and flowing down the undulating hillside.

The site is divided by the main access road, with the upper half hosting the serviced apartments and the lower half featuring a five-star hotel, embraced by landscaped gardens that stretch between the road and the sea. The reception building of the hotel takes a prominent position at the centre of the development, highlighting a spacious two-storey lobby that overlooks the surrounding area. Below the lobby, two floors are dedicated to banqueting and conference facilities, providing separate circulation options as needed. Three specialty restaurants and a pool lounge offer beachfront dining with panoramic views of the resort for guests and residents.

The hotel accommodates 235 rooms in two wings, accompanied by 22 villas located on the westernmost edge of the development. On the northern side of the site, serviced apartments are arranged over nine storeys, with 10 privately owned villas to the east, providing direct access to the beach and hotel facilities. The overall masterplan is seamlessly connected through a network of bridges and pathways.

泰康金融中心

中国，武汉
2015年（建设中）
泰康保险集团股份有限公司

泰康保险集团股份有限公司成立于1996年，现已发展成为中国的保险、资产管理、健康和养老服务行业龙头企业之一。ZHA设计的新泰康金融中心（塔楼）将发扬其卓越的品质，开发高效的服务系统和网络，为各年龄段的人群提供全新的服务与支持。本项目三座塔楼建筑面积共计266,000平方米，现已开始施工，计划于2025年完工。

武汉是中国信息和交通网络中心，泰康金融中心的设计正是以此为出发点。作为汉口滨江商务区城市总体规划中的一部分，三座相连的塔楼形成了一个环抱的圆形组合，将东部的长江公园和湿地以及南部的新中央公园融入其中，宛如跃出江面的江豚。

三座塔楼各自享有绝佳的视野：塔1坐拥城市景观；塔2面向长江；塔3则俯瞰公园。一座花园庭院在三座塔楼之间垂直延伸，宛若一座城市峡谷。在连接高层的空中天桥上设有更多公共空间和设施，如屋顶花园露台，可俯瞰城市和长江河谷的全景。从地面的中央庭院向上延伸至空中天桥露台及环绕中央峡谷的屋顶花园，设计在多个丰富的维度上为城市创造了一个如万花筒般瑰丽的公共空间。泰康金融中心以获得绿色建筑LEED金级认证为标准，三座塔楼呈扇形布局，形成了有效的全天候自遮阳系统。单元式幕墙框体侧面延伸出的楔形飞翼，在为建筑立面提供遮阳的同时保持了室内视野的通透。

Taikang Financial Centre

Wuhan, China
2015 (Under Construction)
Taikang Wuhan

Founded in 1996, Taikang Insurance Group has grown to become one of the largest providers of insurance, asset management, health and elderly care in China. The new Taikang Financial Centre will serve as a centre for leading professionals, working with civic, academic, and corporate organisations. Their collaborative efforts aim to develop effective systems and networks that will create a supportive ecosystem benefitting people of all ages across China. Construction of the three towers within the 266,000-square metre development have now begun, with the centre planned for completion in 2025.

Informed by Wuhan's central role in China's information and transport networks, the design features three interconnected towers situated within the Hankou Riverside Business District. These towers integrate with the Yangtze River park and wetlands to the east as well as the city's new Central Park to the south. Both parks are integral to Wuhan's sponge city programme, facilitating natural rainwater storage and infiltration, flood prevention, and water reuse.

Tower One provides city views, Tower Two faces the river, and Tower Three offers vistas of the park, while a garden courtyard serves as a vertical urban canyon between the towers. Sky bridges connect the towers at higher floors, offering additional public spaces and amenities, including a rooftop garden terrace with panoramic views. Targeting LEED Gold certification, the circular composition ensures effective self-shading throughout the day. External fins with tapered profiles extend from each glazing mullion for additional façade shading whilst maintaining unobstructed views of the city. Leveraging the new technologies developed within the Taikang Financial Centre, the group remains committed to supporting community development, healthcare, education and wellbeing throughout the country.

珠海金湾艺术中心

中国，珠海
2017—2023年
珠海华金开发建设有限公司

珠海金湾艺术中心位于珠海市金湾区，地处西部生态新区的核心地带，旨在成为该地区的当代创意中心。艺术中心与珠机城际铁路相接，与市中心、机场和横琴区的交通联系十分便利。该项目集成了四个文化空间：一个1200座的大剧院、一个500座的多功能厅、一个科普馆，以及一个艺术馆。场馆内部各具特色，以营造独特的到访体验。不同功能空间被整合在一个东西向长170米、南北向长270米连续起伏的整体造型中。

四个文化场馆，两大两小对称排列，并通过中央广场彼此联系，同时围合出一个共享的室外中庭。大剧院和艺术馆采用浅色调的内装材料，而多功能厅和科普馆则选用深色调的内装材质，进一步彰显它们各自的独特个性。

该艺术中心独特的屋顶结构，通过错综复杂的网状壳体将四个场馆连接起来，其灵感来源于候鸟结队飞往中国南方过冬时的"人"字形编队。屋顶网格状钢结构通过重复、对称以及尺度调整，有效地适应了每栋建筑的功能需求。建筑单体的屋顶被优化为大面积可重复使用的预制模块，经过预组装和模块化的施工，均能实现自支撑且具有稳定性。

Zhuhai Jinwan Civic Art Centre

Zhuhai, China
2017-2023
Zhuhai Huajin Development
and Construction

Located at the heart of Jinwan district's Western Ecological New Town, the Zhuhai Jinwan Civic Art Centre is designed as a hub of contemporary creativity for the region. The Art Centre is directly connected to the Zhuhai Airport Intercity railway and has great accessibility to Zhuhai's city centre, the airport, and the Hengqin district. Integrating four distinct cultural institutions for the city, the Art Centre consists of a 1200-seat Grand Theatre, a 500-seat Multifunctional Hall, a Science Centre, and an Art Museum. While each venue offers unique features and captivating visitor experiences, they are unified by a cohesive structure that spans 170 metres wide from east to west and 270 metres long from north to south.

Arranged in a symmetrical layout, the two larger and two smaller venues are interconnected by a central plaza that functions as a shared external foyer for all four cultural institutions. With glazed walls facing the courtyard, visitors can easily differentiate between the individuality and character of each venue, The Grand Theatre and Art Museum share a light materiality, while the Multifunctional Hall and Science Centre exhibit a darker palette of materials, further enhancing their distinct character.

The distinctive roof structure of the Centre connects all four venues through an intricate network of reticulated shells, inspired by the chevron patterns of migratory birds in southern China. These latticed steel canopies, configured through repetition, symmetry and scale variation, effectively meet the diverse functional needs of the institutions. This repetition of the modules optimises pre-fabrication, pre-assembly and modular construction processes, with each building-module of the roof being self-supporting and self-stabilising.

独角兽岛城市设计

中国，成都
2018年（建设中）
成都天府新区投资集团有限公司

成都已处于全球计算机芯片生产的领先地位，不仅为全球一半的笔记本电脑提供芯片，还是移动计算机硬件的重要生产地。位于成都天府新区的独角兽岛城市设计占地67公顷，将进一步推动中国数字经济的发展，为国内外各类公司提供优质的居住和办公环境。

独角兽岛城市设计是一项综合性的规划方案，旨在为7万名研究人员、公司职员、居民及访客提供服务，增强社区环境体验。该地区的都江堰灌溉系统建造于2300年前，已被联合国教科文组织列入"世界文化遗产名录"。这一古老的水利设施借助河流的自然力量为成都平原提供灌溉，不仅确保了农业的繁荣，还保证了该地区的防洪能力。受这一历史悠久的水利工程的启发，独角兽岛的设计结合了都江堰灌溉系统的环保原则，将绿色城市空间、节水设施及连通便利性结合起来，创造可持续的居住和办公环境。

该岛的设计围绕中央广场展开，形成连接地铁站和地下停车设施的交通网络，保证在方圆600米范围内，可以通过步行或短途自行车轻松抵达。中央广场作为主要的聚集场所，提供多种室内外活动空间，使下沉广场充满活力。这片开阔的区域还可以承载展示和推广新产品的功能，支持举办小型商业活动。

Unicorn Island Masterplan

Chengdu, China
2018 (Under Construction)
TIANFU Group, Chengdu

Chengdu has emerged as a global leader in computer chip production, supplying half of the world's laptops, and is a prominent manufacturer of mobile computing hardware. Located in the Tianfu New Area, the 67-hectare Unicorn Island masterplan aims to further propel China's digital economy by creating vibrant living and working environments for both domestic and international companies.

Developed as a mixed-use masterplan for 70,000 researchers, office staff, residents, and visitors, Unicorn Island is designed to enhance the wellbeing of its community. Drawing inspiration from the region's historical natural engineering projects, the design of the masterplan incorporates principles derived from the Dujiangyan irrigation system, a UNESCO world heritage site dating back 2,300 years. This ancient water management system harnessed the natural forces of the Min river to irrigate the Chengdu plain, mitigating flooding and ensuring agricultural prosperity. Guided by these environmental principles, the design of the masterplan integrates green civic spaces, water conservation measures, and enhanced connectivity to cultivate sustainable living and working environments.

The Island is centred around a main plaza, serving as a hub for connectivity with its metro station and underground parking facility. Within a radius of 600 metres, all parts of the island are easily accessible through a short walk or bike ride, ensuring convenient transportation. The central plaza acts as the main gathering place, offering a variety of outdoor and indoor spaces that animate the underground concourse. This expansive area provides opportunities to showcase and promote new products, and accommodates light commercial activities.

恒基地产大厦

中国，香港
2018（建设中）
恒基兆业地产集团

恒基地产大厦位于香港中央商务区的中心，这里曾是一个多层停车场。大厦共36层，旨在打造一片城市绿洲。场地邻近遮打花园，步行即可到达中环与金钟地铁站。塔楼的裙楼为下方栽有绿植的庭院花园提供遮蔽，也为世界最繁忙的城市之一——香港的中心地带提供了一个被大自然包围的新城市广场。

建筑呼应大自然的有机形式，并与邻近的公共花园相连。塔楼室外区域的静谧被引入内部宽敞的公共空间。以高超的制作工艺和卓越的施工精度打造的弧形玻璃幕墙增强了建筑内部与周边花园和城市之间的无缝连接。该设计诠释了含苞待放的紫荆花的结构形式与层次感——作为香港特别行政区的区花，紫荆花最早由美利道地块上的植物园培植，并出现在香港特别行政区的区旗上。

塔楼的大部分空间被设计为智能化办公空间，优先考虑用户的舒适与健康。建筑采用了创新技术，如空气清洁系统、智能免接触系统、人工智能电梯控制系统（以减少用户等待时间），以及隔热幕墙（以过滤多余热量）。高强度钢结构塑造了可以提供更多自然采光且宽敞、舒适的甲级办公空间，无柱跨度宽达26米，层高达5米，确保了最佳的灵活性。通过ZHA、恒基地产和ARUP的合作，这项设计获得了LEED白金与WELL白金预认证，以及中国绿色建筑的最高等级绿色三星评级。

The Henderson

Hong Kong, China
2018 (Under Construction)
Henderson Land

Located in the heart of Hong Kong's central business district, 'The Henderson' tower for Henderson Land stands at 36 storeys, replacing a multi-storey car park to create an urban oasis. Its location adjacent to Chater Garden is within a short walking distance to both Central and Admiralty MTR metro stations. The tower's base is elevated above the ground to shelter courtyards and gardens cultivated with trees and plants, creating new civic plazas enveloped by nature in the centre of one of the world's busiest cities.

This redevelopment project echoes the organic forms of the natural world and connects with the neighbouring public gardens and parks. The serene outdoor areas of the tower flow into the spacious communal spaces of the interior. The craftsmanship and precision of the curved glass façade enhance this seamless connectivity between the building's interiors and the surrounding gardens and cityscape. The design reinterprets the structural forms and layering of a Bauhinia bud about to blossom. Known as the Hong Kong orchid tree, the Bauhinia blakeana was first propagated in the city's botanic gardens above the Murray Road site, and its petal is featured on Hong Kong's flag.

Much of the tower is designed as smart offices and workplaces, prioritising user comfort and wellbeing. It incorporates innovative technologies, such as clean air systems, smart contactless journeys for tenants, AI-optimised lift control to minimise wait times, and a thermal-insulating façade to filter excessive heat. The building features a high-tensile steel structure, enabling expansive and naturally illuminated Grade A office spaces with column-free spans of up to 26 metres and a generous floor-to-floor height of 5 metres, ensuring optimal flexibility. Through collaboration with Henderson Land and Arup, the design has achieved LEED Platinum and WELL Platinum pre-certification as well as the highest 3-Star rating of China's Green Building Rating Program.

虹桥文化中心

中国，上海
2018年（未建成）
虹桥镇文化中心

虹桥文化中心地处长宁区，这里是上海一个充满活力的创新中心。项目位置邻近以教育、技术和文化景观而闻名的静安区。ZHA在设计中完美地结合灵活性、流动性、娱乐性和社区精神，为虹桥文化中心项目创造了一座强调社区活力的"集体城市"。设计贴合社区、科技、教育和体育规划等不同功能的多种需求，形成了一个和谐共生的建筑方案。

建筑组合了多个简单体块，根据特定的视线方位和日光分析成果来有序定位，让人联想到中国传统园林中的天井。这种排布方式促进了多层次功能整合，并以局部挑空为点缀。挑空区域周边的丰富空间，以科学、文化和体育等组合的形式鼓励各类用户之间的互动交流；在为虹桥文化中心项目打造统一风格的同时，最大限度提高灵活性、效率和未来适应性。

与通常在城市场地中出现的传统线性、垂直布局不同，该建筑的设计规划沿环境轴线按不同功能进行分层，曾经孤立的功能如今汇聚重叠，摇身变为全新的社区中心。内部交流区和露台穿插在整个场地之中，这一组合增强了文化中心的内部连通性，同时培养、促进了社区意识。

Hongqiao Cultural Centre

Shanghai, China
2018 (Unbuilt)
Hongqiao Cultural Centre

Hongqiao Cultural Centre is situated in the Changning district, a vibrant hub for innovation in Shanghai. It strategically stands near Jingan District, which is renowned for its thriving education, technology, and cultural scene. The conceptual design proposed by Zaha Hadid Architects embodies the notion of a 'collective city,' accentuating the dynamism of community life through the integration of flexibility, fluidity, entertainment, and communal engagement. The design adeptly accommodates the diverse requirements and functions of the centre, including Community, Science and Technology, Education, as well as Sports programmes, resulting in a harmonious and responsive architectural solution.

The organisation of the architecture comprises a series of simple volumes, positioned in response to contextual vista axes and sunlight analyses. Reminiscent of the light-courts observed in traditional Chinese gardens, this deliberate placement fosters a porous integration of functions, punctuated by striking voids. Encircling these voids, a multi-layered arrangement of programs stimulates interactive spaces between different users, creating a dynamic operational assembly of science, culture and sport. This approach optimises for flexibility, efficiency, and future adaptability, while establishing a cohesive identity for the architecture.

Departing from the conventional linear and vertical layout typically found in urban sites, the design of the architecture stratifies diverse functions along contextually determined axes. Previously isolated elements are now overlapped and converged, forming a new community hub with internal communication areas and terraces interspersed throughout the site. This arrangement encourages connectivity and fosters a sense of community within the Cultural Centre.

西安腾讯双创小镇城市设计

中国，西安
2018年（未建成）
迈科投资控股有限公司

西安腾讯双创小镇旨在用创新和富有想象力的设计来满足用户需求，创造一个商业服务、基础设施和休闲活动无缝衔接的综合体，营造一处相互连接、对行人友好的无障碍环境。

小镇的设计充分响应环境，通过补充公共空间来承载更多社区活动，场地内各类设施有多种使用方式。用途的多样性将增强商业区和服务区功能。高度参与感和幸福感是小镇设计策略的关键所在。

腾讯双创小镇同时提供卓越的住宅产品，兼具功能性、定制化服务和社区氛围。整体设计规划遵循紧凑型城市框架的原则，强调高效的土地利用和连通便捷性。该小镇以开放、包容的姿态提供了集生活、工作和文化于一体的创新场所。受中国传统庭院建筑的启发，住宅群围绕着裙楼内部的室外庭院展开，中央主干道与所有庭院之间的连接使交通更加便利，也更具有社区氛围。

Xi'an Tencent Innovation City Masterplan

Xi'an, China
2018 (Unbuilt)
Maike Investment Holdings Group

The Tencent Innovation City Masterplan is designed to meet the needs of its users in an innovative and imaginative way. It aims to create a seamless integration of services, infrastructure, and leisure activities, fostering an interconnected, pedestrian-friendly, and accessible environment.

The city will be a fully responsive environment, offering multiple ways to engage with its amenities, complementing rich public spaces with community-based activities. Diversity of uses will enhance commercial and service areas where richness of experience and well-being are key components of the urban strategy.

Tencent Innovation City aims to provide exceptional residential offerings that balance functionality, exclusivity, and community. The masterplan adheres to the principles of the Compact City Framework, which emphasises efficient land use and connectivity. The city will serve as an inclusive home, workplace, and cultural hub. Inspired by the traditional Chinese courtyard building typology, the residential clusters are structured around a podium with a central courtyard. The direct connectivity between the central spine and all courtyards promotes a strong sense of community and facilitates connections.

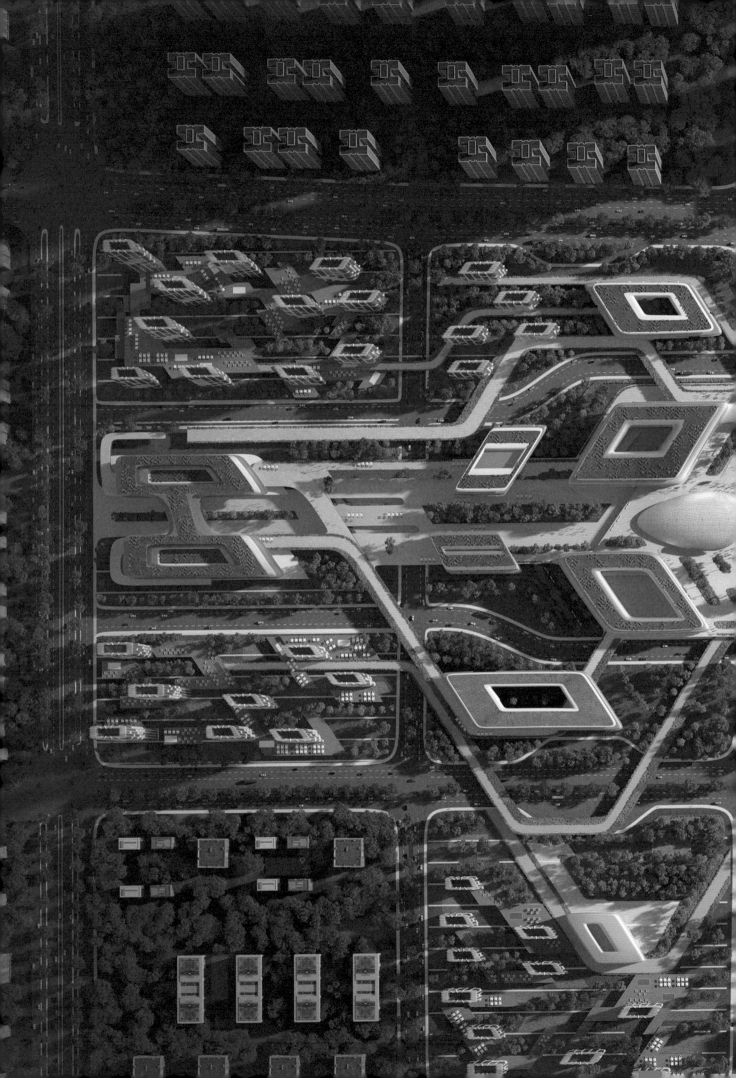

杭州江河汇无缝城市
杭州国际金融中心

中国，杭州
2018年（建设中）
新鸿基地产发展有限公司
平安不动产有限公司
钱城开发

杭州国际金融中心地处钱江新城核心区域，地理位置优越。该建筑位于钱塘江与京杭大运河交汇处，由ZHA设计的杭州河中部公园环绕。杭州国际金融中心总建筑面积822,000平方米，由多个地块组成，项目设计充分利用了其得天独厚的地理位置。它蕴含着杭州丰富的历史底蕴和光明灿烂的未来，展示了杭州具有地域特色的非物质文化遗产。

杭州国际金融中心的设计理念源于非物质文化遗产"蚕丝织造技艺（杭罗织造技艺）"。杭罗产自杭州，历史悠久，可追溯到宋代。杭罗织造技艺作为中国传统桑蚕丝织技艺的重要子项目之一，于2009年被列入联合国教科文组织"人类非物质文化遗产代表作名录"。

这一精心打造的综合体完美结合了各种设施，包括高端住宅、现代化办公空间、充满活力的购物中心、高档酒店，以及便捷的服务式公寓。

Seamless City: Hangzhou Qianjiang New Town River Confluence Masterplan
Hangzhou International Financial Centre

Hangzhou, China
2018 (Under Construction)
Sun Hung Kai Properties Ltd +
Ping An Real Estate +
Qiancheng Development

Hangzhou International Finance Center (Hangzhou IFC) is a strategically located architectural development in the core area of Qianjiang New City. It is surrounded by the new Hangzhou River Middle Park, designed by ZHA, at the confluence of the Qiantang River and the Beijing-Hangzhou Grand Canal. Spanning an impressive 822,000 square metres, this expansive development is composed of multiple plots, taking advantage of its unique geographical location. It embraces the city's rich history and promising future, showcasing Hangzhou's distinctive intangible cultural heritage.

The design concept of this comprehensive development draws inspiration from the intricate art of 'Hangluo weaving skills.' Hangluo, produced in Hangzhou, has a long history dating back to ancient times. It represents the traditional Chinese mulberry silk weaving technique and was recognized by UNESCO as part of the 'Representative List of Human Intangible Cultural Heritage' in 2009.

The masterfully crafted complex seamlessly integrates a wide range of high-quality amenities, convenient serviced apartments.

杭州江河汇无缝城市
杭州河中部公园：
景芳三堡单元江河汇城市综合体汇中区块项目

中国，杭州
2018年（建设中）
杭州市钱江新城开发集团有限公司

杭州河中部公园位于钱塘江与京杭大运河交汇处。作为河流交汇处的观潮点，游客在这里就可以观赏到潮涨潮落。

公园内的一座以"杭罗织造技艺"为建筑设计灵感的人行天桥和一座150米高的摩天轮格外引人注目，游客在摩天轮上可以俯瞰整个区域。这些元素与无缝整合的设施和全景绿化露台相辅相成，为公园增添了独特的魅力。

公园的设计理念围绕京杭运河展开，以横跨运河的物理和视觉连接为象征，加强不同地块与建筑之间的凝聚力。

公园分为西堤和东堤，由天桥连接。西侧设有中央广场，作为城市门户，通过一条人行高架道路连接南侧的南部购物村与北侧的地铁站。东侧的主要设施位于现有泵站的区域内，该区域将通过增设屋顶花园和露台来进行改造，以提供优美的河流与运河视野。这些设施将包括游客中心、文化设施与摩天轮专用空间。

公园与人行天桥占地约14.7公顷，当下正在建设中，巨型摩天轮目前也正在进行规划审批，它将为这一地标性建筑带来振奋人心的未来。

Seamless City: Hangzhou Qianjiang New Town River Confluence Masterplan
Hangzhou River Middle Park: Jingfang Sanbao Unit Jianghehui City Complex Huizhong Block Project

Hangzhou, China
2018 (Under Construction)
HANGZHOU CBD Development Group Co., Ltd

The River Middle Park in Hangzhou is situated at the confluence of the Qiantang River and the Beijing-Hangzhou Grand Canal, known locally as the Jing—Hang Grand Canal. As a river-confluence tidal observation point, it allows visitors to witness the natural ebb and flow of the tides.

Notable features of the park include a footbridge inspired by Hangluo weaving skills and a 150-metre tall ferris wheel offering elevated views of the region. These elements contribute to the park's unique charm, complemented by seamlessly integrated facilities and panoramic green terraces.

The park's design concept revolves around the interconnectedness of the Jing—Hang Canal, symbolised by physical and visual connections spanning across the canal and fostering cohesion among different plots and buildings.

Divided into West and East embankments, the park is connected by the footbridge. The West side features a Central Plaza, serving as an urban gateway that links a southern shopping village to the South and the MTR station to the North via a pedestrian elevated path. On the East side, main facilities are located within the area of an existing pump station, which will be transformed with the addition of roof gardens and terraces, providing captivating views of the river and canal. These facilities will include a visitors centre, cultural facilities, and spaces dedicated to the ferris wheel.

Covering an approximate area of 14.7 hectares, the park and footbridge are under construction, and the giant ferris wheel, currently undergoing planning approval, is promising an exciting future for this landmark project.

OPPO 国际总部

中国，深圳
2020年（建设中）
OPPO广东移动通信有限公司

作为中国领先的智能手机制造商，OPPO公司是智能设备和互联网服务领域通信技术革新的先锋。OPPO国际总部将成为OPPO品牌与理念的有力体现，反映出OPPO公司对技术艺术性、永恒之美，以及与自然和谐融合的承诺，更将成为城市的重要资产。

新总部由四座彼此相连的塔楼组成，建筑高度200米（40层），总面积为185,000平方米。其中，两座塔楼采用灵活的开放式布局，并通过一个14层高的竖向中庭连接起来；另外两座塔楼作为服务空间，提供了竖向的交通流线。塔楼的朝向设计特别考虑要最大限度地保留深圳湾的景观。塔楼在低层逐渐内缩，为低层留出了更多公共空间。

考虑到项目的空间复杂性，ZHA采用三维建筑信息模型与能源管理系统以优化建筑效率。由于大楼的核心筒被偏置在两栋服务塔楼中，因此另外两座功能塔楼内部拥有不受阻挡的开阔视野，方便员工之间互动。中部的竖向中庭在视觉上将内部各个空间联系了起来，有助于公司不同部门之间的协作。该设计通过充足的自然光、多样的工作环境，以及工作人员与访客多样化的交通流线，为创造性的参与和自发性的活动提供了有利的环境。

OPPO Headquarters

Shenzhen, China
2020 (Under Construction)
Oppo Telecommunications

As China's leading smartphone manufacturer, OPPO pioneers communication technology innovations in smart devices and internet services. Currently under construction, the new OPPO Headquarters in Shenzhen will serve as a powerful architectural embodiment of OPPO's brand and philosophy, reflecting their commitment to technological artistry, timeless beauty, and harmonious integration with nature, thereby positioning it as a significant asset for the city.

The design of the Headquarters comprises four interconnected towers, with a height of 200 metres (40 floors) and a total area of 185,000 square metres. Two of the towers feature flexible and open-plan spaces, connected by a 14-storey vertical lobby, while the other two towers serve as external service towers for vertical circulation. With a strategic orientation to maximise the scenic views of Shenzhen Bay, the towers gradually taper inwards at lower levels, creating expansive civic spaces at street level.

The project has been developed with 3D Building Information Modelling and energy management systems to optimise efficiencies. By locating the service cores of the towers externally, the interior of each floor remains unobstructed, facilitating uninterrupted interaction among employees. The incorporation of spacious atriums ensures visual connectivity, fostering collaboration between various departments within the Headquarters. The design promotes a favourable environment for creative engagement and spontaneity through the abundance of natural light, varied work settings, and a diverse circulation path for staff and visitors.

深圳金融科技研究院

中国，深圳
2020年（建设中）
深圳市福田区建筑工务署

深圳金融科技研究院位于福田中央商务区，紧邻深圳中央公园和深圳会展中心。这座200米高的塔楼将与大百汇中心和平安国际金融中心共同勾勒出福田区的天际线。

塔楼体量是根据场地限制精心推敲而成的，建筑外壳的几何造型也考虑了日照限制及与相邻建筑的关系。立面设计语言以视觉方式再现了深圳金融科技研究院的核心功能——"前沿、科技、律动"的设计理念，捕捉到了金融数字化创新的瞬息性。幕墙及室内设计风格结合了数字货币的高效流动、快速交易等特点及区块链等创新技术，体现了去中心化、全方位的流动性和连通性的概念。立面上流动转折的图形，围合并重组建筑空间，形成一种看似持续变化的状态。

建筑围护结构的隔热性能和塔楼幕墙设计的变化处理，增强了建筑的节能效果。根据日照分析，随着塔楼高度变化，立面采用了不同的玻璃类型和百叶密度。空调、照明、电梯及控制设备都采用了特定的节能调控方式。高性能给排水设备，结合高效灌溉方式，能够达到当地的节水标准。选用高强度钢减少了建筑中的总体用钢量。郁郁葱葱的屋顶花园和宽敞的内部空间增强了用户的幸福感。

Shenzhen Institute for Financial Technology

Shenzhen, China
2020 (Under Construction)
Bureau of Public Works of
Shenzhen Futian Municipality

The Tower of Shenzhen Institute for Financial Technology is located in the Central Business District of Futian, in close proximity to the Shenzhen Central Park and the Shenzhen Convention and Exhibition Centre. The new 200-metre tall tower shares the Futian skyline with Dabaihui Centre and the Ping'an International Financial Centre.

The tower massing is carefully adapted to the site conditions, with the envelope carved to the sunlight constraints of the site with respect to the neighbouring buildings. The design for the new tower embodies the core activities of the Shenzhen Institute for Financial Technology. The 'High Tech Flow' design concept captures the ephemeral qualities of digital innovation in finance. The dynamic façade and interiors reference the efficient flow of digital currency, fast transaction speeds, and innovative technologies such as blockchain. It embodies the concepts of decentralisation, all-encompassing fluidity, and connectivity, capturing the essence of online network-based dynamics. Flow patterns wrap and organise the building, resulting in a seemingly constant state of transformation.

Energy saving is achieved by thermal insulation of the building envelope and treatment of the tower façade. Informed by sun exposure analysis, different glazing types and varying louvre densities are integrated into the building skin along the height of the tower. Specific energy-saving applications are adopted for air conditioning, lighting equipment, elevators and control measures. High-performance water supply and drainage equipment is used in combination with high-efficiency appliances and irrigation methods to meet local water-saving standards. High-strength steel is selected to reduce the overall amount of steel used in the building. A luscious rooftop garden and spacious internal voids enhance the well-being of its occupants.

The New World
Zaha Hadid Architects

中节能·上海首座项目

中国，上海
2020年（建设中）
中国节能环保集团有限公司

中国节能环保集团是中国节能环保行业的领军企业。中节能·上海首座项目的设计将展示其对可再生能源与环境保护始终如一的承诺。园区毗邻黄浦江上的杨浦大桥，以上海浓厚的历史为背景，该项目旨在成为上海最环保的可持续性建筑。由于可持续性渗透了项目设计与建造的各个环节，该建筑以超高评分获得了中国绿色建筑三星认证。建筑面积218,000平方米的中节能·上海首座将在节能、能效及可持续性方面树立行业新标杆，开创城市先河。

该项目由三座办公塔楼与集购物、餐饮和休闲功能于一体的公共设施组成，并借助城市景观广场与周边城市社区和黄浦江连接起来。项目由一系列环环相扣的连锁结构组成，以控制建筑在尺度上的压迫感。塔楼设置了空中露台，为使用者提供室外休憩的场所。此外，该项目还涉及对30年代的历史建筑明华糖厂的复兴和重新规划，以容纳中节能·上海首座项目多样化的教育和社区参与项目。

利用ZHA开发的先进算法平台，该项目优化了建筑体量、朝向和幕墙与楼层的比例，整合了被动式设计原则，以减少能耗并尽量降低太阳能得热，优化的外部遮阳系统有效降低了内部的制冷需求。此外，该项目还将包含地源热泵技术、中国节能环保集团开发的光伏及可循环材料、雨水回收等可持续技术，同时利用景观水体以在夏季创造出凉爽的场地微环境。

CECEP Shanghai

Shanghai, China
2020 (Under Construction)
China Energy Conservation
and Environmental Protection Group

CECEP, China's leading energy conservation and environmental protection enterprise, demonstrates its ongoing commitment to renewable energy and environmental stewardship through the design of its new Shanghai campus. Situated adjacent to the Yangpu Bridge on the Huangpu River and defined by Shanghai's vibrant history, the campus aims to be the city's most environmentally sustainable building. With sustainability embedded into every aspect of its design and construction, the building achieved over 90 credits in China's Three Star Green Building Rating System. The 218,000-square metre campus establishes new standards in energy conservation, efficiency, and sustainability, setting a precedent for the city.

The mixed-use urban campus, consisting of three office towers, shopping, dining and leisure facilities, is linked together by a park that connects directly with the city and river. Composed as a series of interlocking rings that reduce its perceived scale, the design incorporates external sky terrace that provide public spaces and connect exterior and interior spaces. Moreover, the project involves the revitalisation and repurposing of the historic Minghua Sugar Factory, a 1930 industrial building, to accommodate CECEP's diverse educational and community engagement programmes.

Utilising advanced modelling tools developed by ZHA, the project optimises architectural massing, orientation, and façade-to-floor ratios to integrate passive design principles, reducing energy consumption and minimising solar heat gain. The design employs optimised external shading to minimise cooling demands. Furthermore, the design incorporates sustainable technologies such as Ground-Source Heat Pumps (GSHP), photovoltaic and recyclable materials developed by CECEP, and rainwater harvesting. The design also uses integrating water features in the landscape to create a comfortable micro-climate in summer.

皇岗口岸区域规划

中国，深圳
2020年（进行中）
深圳深港科技创新合作区发展有限公司

深圳皇岗口岸区域规划将成为广深科技走廊的重要节点。皇岗口岸占地1.67平方公里，是深港铁路与公路全天候通达的门户，每天约有30万人次由此通过。经过全面升级和改造，该项目不仅能成为连接大湾区的关键枢纽，还将成为国际技术创新的平台，促进微电子、材料开发、人工智能、机器人技术和医学等产业的科研与联合。

皇岗口岸区域规划以两个大型公共广场为中心，界定了三个相互联系的区域：港口枢纽区、协同创新区和港口居住区。港口枢纽区位于规划的西南部，设有行政办公室和酒店，将供来访科学家和研究人员使用。协同创新区由原有的停车场和货检区改造而成，由科研中心、实验室和会议中心组成，促进知识交流与合作。港口居住区位于东北部地区，邻近地铁站，将涵盖住宅、学校、体育和娱乐设施，以及一系列购物与餐饮设施。

基于现有的街道布局，规划的主轴线发展成为宽敞的人行大道，连接两个主要公共空间：东北部地铁站旁的市民广场与西南部行政中心的大型公共广场。放射状的道路进一步将区域划分成不同的组团，并通过共享平台及天桥将各区域连接起来，创造出功能性附属流线，并适应研究机构在未来的扩展。此外，每个建筑组团都有自己的核心广场，确保将自然环境融入区域规划。

Huanggang Port Area Masterplan

Shenzhen, China
2020 (Ongoing)
Shenzhen—Hong Kong Science and Technology Innovation Cooperation Zone Development Co., Ltd.

The Huanggang Port Area Masterplan in Shenzhen will serve as a vital node within the Guangzhou-Shenzhen Science and Technology Corridor. Spanning 1.67 square kilometres and operating as a 24/7 portal for Hong Kong—Shenzhen railways and highways, Huanggang Port facilitates the daily passage of approximately 300,000 individuals across the border. Through its comprehensive upgrading and reconstruction, the New Huanggang Port will not only act as a hub for the linkage of the Greater Bay Area, but also serve as an international platform for technological innovation, fostering scientific research and collaboration across diverse industries such as microelectronics, material development, artificial intelligence, robotics, and medical sciences.

Structured around two primary public plazas, the masterplan consists of three interconnected districts: the port hub, the collaborative innovation area, and the port living zone. The port hub district will be situated in the southwestern section and will feature administrative offices and hotels for visiting scientists and researchers. The collaborative innovation area, transformed from former parking lots and cargo inspection areas, will house scientific research centres, laboratories, and a conference centre, facilitating knowledge exchange and collaboration. The port living district, located near the metro station in the northeastern area, will include residential buildings, schools, sports facilities, recreational amenities, as well as a range of retail and dining establishments.

Drawing from the existing street layout of Shenzhen, the masterplan establishes a base grid, with a spacious pedestrian boulevard serving as the central spine, connecting two primary public gathering spaces: the civic plaza at Fulin metro station in the northeast and the expansive public square at the heart of the administrative hub in the southwest. A radial secondary grid further subdivides the districts into clusters, creating additional circulation routes to enhance functionality and accommodate future expansions of research facilities through shared podiums and bridging skywalks. Additionally, each smaller cluster of buildings incorporates its own 'nucleus' of an outdoor communal square, ensuring the integration of natural spaces throughout the masterplan.

深圳湾超级总部基地C塔

中国，深圳
2020年（进行中）
深圳湾超级总部基地

深圳湾超级总部基地是粤港澳大湾区重要的商业和金融中心，可容纳企业总部及约300,000名员工，C塔位于规划的城市绿轴与城市走廊交会处，地理位置十分重要。该设计将塔楼附近的公园与广场以阶梯景观的形式引入场地并向上延伸至两座塔楼内部，使塔楼完美融入周边环境，从而吸引公众进入建筑中心。连接两座塔楼的环形连桥设有文化与休闲场所，吸引公众驻足流连并欣赏城市的全景。

ZHA还借助三维建模工具对C塔的建筑体量、朝向及窗墙比例关系进行了优化。该项目由两座约400米高的塔楼组成，形成了一个多维度垂直都市，提供无柱且可以自然采光的办公、购物、娱乐及餐饮空间，同时配以酒店、会议中心及带有展览空间的文化设施。

C塔采用双层隔热单元式玻璃幕墙，其垂直的通道可提供自遮阳与通风调节，通过自然与混合通风确保实现高效的环境控制。塔楼的设计通过与该区域的智能管理系统相连，持续监测内外部环境，同时利用先进的室内环境控制技术、水收集与循环系统以及太阳能光伏板系统来优化能源使用。每层的露台都设有水景花园来对当地环境中的污染物进行生物过滤，同时使用低挥发性的有机化合物材料来尽可能地减少室内污染物和微粒，以保持室内空气质量。

Tower C at Shenzhen Bay Super Headquarters Base

Shenzhen, China
2020 (Ongoing)
Shenzhen Bay Super Base
Construction Headquarters Office

The Shenzhen Bay Super Headquarters Base represents a significant business and financial hub that will serve the Greater Bay Area of Guangdong, Hong Kong, and Macau. Designed to accommodate corporate headquarters and approximately 300,000 employees daily, the development holds a prominent position at the intersection of the city's planned green axis and urban corridor. Tower C seamlessly integrates with its surroundings, forming a harmonious connection with the adjacent park and plazas. The design transforms these spaces into a terraced landscape that extends vertically within the two towers. Open and inviting, the building welcomes the public inside. Cultural and leisure attractions housed in sweeping bridges that link the towers draw people in and offer panoramic views of the cityscape.

Utilising 3D modelling tools developed by ZHA, the design of Tower C optimises efficiencies in architectural massing, orientation, and façade-to-floor ratios. Comprising two towers at nearly 400 metres, the building forms a multi-dimensional vertical city, providing column-free and naturally lit office space. It houses a range of amenities including shopping, entertainment, and dining amenities, and it also features a hotel, convention centre, and cultural facilities with exhibition galleries.

Tower C's design incorporates a double-insulated, unitised glass curtain wall with vertical channels that provide self-shading and ventilating registers, ensuring efficient environmental control on each floor through natural and hybrid ventilation. Seamlessly integrating with the district's smart management systems, the building constantly monitors external and interior conditions, while incorporating advanced indoor environmental controls, water-collection and recycling systems, and photovoltaics for solar energy harvesting, all aimed at reducing energy consumption. The inclusion of aquaponics gardens on each terrace level provides biological filtration of local contaminants, while the use of low-volatile organic compound materials helps maintain indoor air quality by minimising indoor pollutants and particulates.

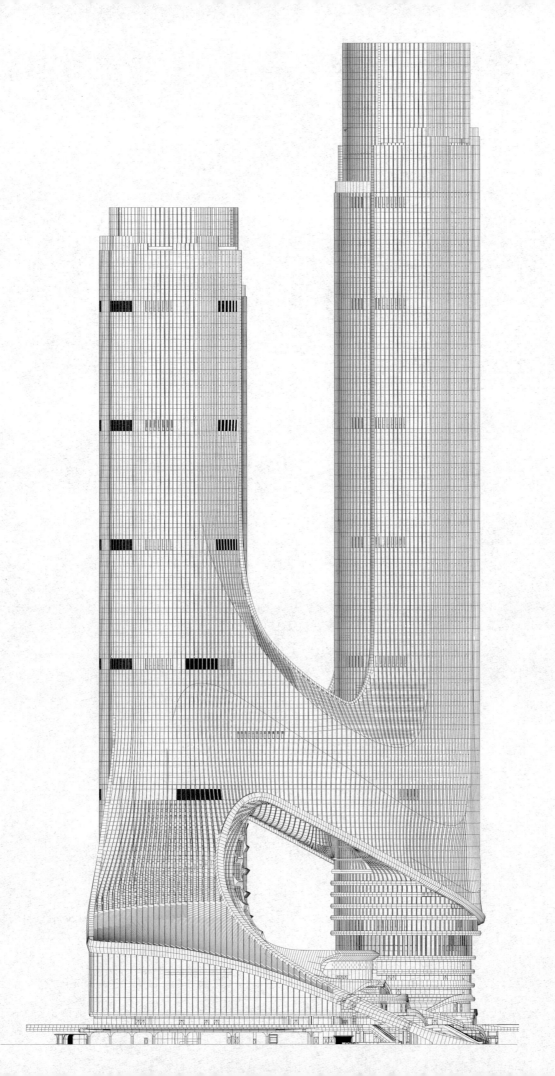

杭州未来科技文化中心国际体育中心

中国，杭州
2022年（进行中）
浙江杭州未来科技城管理委员会

杭州未来科技文化中心国际体育中心位于杭州未来科技文化区，包括一座可容纳60,000个座位的足球场、两个练习场、一座有19,000个座席的室内体育场，以及一座可提供两个50米标准泳池的水上运动中心。该设计还包括一个全新的滨江公园及公共广场，为通往杭州不断发展的地下交通网络提供了便利。

为适应杭州日益增长的人口，国际体育中心旨在满足广大运动爱好者与专业运动员的各项需求。紧凑的场馆设计，加上合理的朝向布局，使得近一半的场地转变为全新的城市公共空间，并完美融入杭州的城市规划及河岸的自然景观。13.5万平方米的足球场坐落于公园的东侧，面向城市开放。毗邻足球场，室内体育场和水上运动中心通过分层平台相互连接。平台的设计灵感来源于杭州山区的梯田茶园，45,000平方米的平台裙房承载了体育中心的附属设施，包括训练室、更衣室、办公室、商店及能俯瞰庭院与户外平台的餐厅。

与单一用途的传统大型体育场不同，该中心的室内场馆可独立于体育场运作，用于举办各类活动，为运营提供极大的灵活性，从而确保该中心成为活跃的全天候社区集会场所。国际体育中心按照国际足联的标准设计，"碗"状布局使体育场内的观众席尽可能地接近球场，确保全部座席的视野均不受限制，为球员和球迷创造一种热烈的赛时氛围。建筑外立面的横向百叶构件结合室内外功能融合的需求，模糊了建筑内外的边界。

Hangzhou International Sports Centre

Hangzhou, China
2022 (Ongoing)
Zhejiang Hangzhou
Future Science and Technology City
Management Committee

Situated in Hangzhou's Future Science and Technology Cultural District, the Hangzhou International Sports Centre incorporates a 60,000-seat football stadium, two practice pitches, a 19,000-seat indoor arena, and an aquatics centre with two 50-metre pools. The design also establishes a new riverfront park and public plazas, providing convenient access to the city's expanding metro network.

Accommodating Hangzhou's growing population, the centre caters to a range of athletes, from grassroots players to professionals. With a compact design and deliberate orientation, nearly half of the site is dedicated to new parks and gathering areas, enhancing the urban plan and natural landscapes along the riverbank. The 135,000-square metre football stadium is positioned on the eastern side of the new park, facing the city. Adjacent to the stadium, the indoor arena and aquatics centre are connected through a layered podium that spans the site. Inspired by the terraced tea farms on Hangzhou's hillsides, the 45,000-square metre striated podium accommodates shared ancillary facilities such as training halls, locker rooms, offices, shops, and restaurants overlooking the courtyard and terraces.

Unlike conventional large stadiums with single-use programs, the indoor arena at the centre operates independently from the stadium, offering maximum flexibility to host a diverse range of events. This design approach ensures that the centre serves as a vibrant community gathering place throughout the day and evening. The stadium, designed to FIFA standards, features a seating bowl that is configured to bring spectators as close as possible to the field, ensuring unrestricted views from every seat and creating an intense matchday atmosphere for both players and fans. These programmatic requirements define the undulated geometries within the louvred façade, blurring the boundary between the interior and exterior.

泾河新城国际文化艺术中心

中国，西安
2022年（建设中）
陕西省西咸新区泾河新城投资发展有限公司

项目坐落于西安北部的泾河新城，致力于发展成为以新能源新材料、人工智能和航空航天为重点的产业区。泾河新城国际文化艺术中心位于泾河湾院士谷核心区，建筑形态再现了泾河优雅的曲线，如流水雕刻山脉，与周边的景色相辅相成。

该中心的设计回应区域总体规划。高线景观云廊跨越泾河大道，将主干道北侧的多媒体图书馆与南侧的多功能剧场、艺术教育工作室及展馆等设施联系起来。高线云廊的设计消解了城市道路对项目连续性的破坏，平缓的坡道形成了通往景观平台的入口，缝合、串联起场地北侧的商业、居住区与南侧的公园、河流。

建筑由"飘浮流动"的银色金属幕墙及其下的玻璃体块和室外景观平台交织而成。这些元素与景观相互联系，为当地社区定义了一系列室内外文化、休闲空间。多媒体图书馆的室内平台沿中庭向上螺旋攀升，将为个人及团队研究提供各类或安静或开放的阅读区。图书馆结合实体出版物与沉浸式虚拟现实技术，拓宽了学习的边界，让人们能够以更加丰富的形式交流知识。艺术中心剧场位于大道南侧，可容纳1000人，空间形式可灵活变化以适应不同类型的活动。艺术教育工作室和展廊交叠环绕在剧场周围，多个共享公共空间在提升可达性的同时，加强了各种文化活动间的互动。

Jinghe New City Culture and Art Centre

Xi'an, China
2022 (Under Construction)
Shaanxi Xixian New District
Jinghe New Town
Investment Development

The Jinghe New City, situated in the northern part of Xi'an, is emerging as a prominent hub for the development of industries focused on new energy and materials, artificial intelligence, and aerospace. Embracing the natural beauty of the surrounding Shaanxi province, where the Jinghe River gracefully carves through mountains and landscapes, the Jinghe New City Culture and Art Centre finds its place within the Jinghe Bay Academician Science and Technology Innovation district of the city.

The design of the centre seamlessly integrates with the city's existing urban masterplan, establishing a connection between the new multimedia library located to the north of Jinghe Avenue and various facilities to the south, including the performing arts theatre, multi-function halls, studios, and exhibition galleries. This integration is achieved through elevated courtyards, gardens, and pathways that gracefully traverse the eight lanes of traffic below, uniting the two sides of the avenue. Gently sloping ramps serve as gateways to the district's network of elevated public walkways, enabling the centre to weave through the city and connect its commercial and residential districts with the parks and river to the south.

Organised as a series of flowing volumes, layers, and surfaces interconnecting with courtyards and landscapes, the design defines a sequence of interior and exterior cultural and recreational spaces for its community. The multimedia library features terraces overlooking its full-height atrium, offering diverse reading areas for individual and collaborative research. It integrates traditional print publications with immersive virtual reality technologies, expanding the horizons of knowledge exchange. On the southern side of the avenue, the performing arts theatre, with a capacity of 1000 people, provides a versatile venue for various events. Adjacent to the theatre, the multi-function hall, studios, and galleries are strategically arranged to promote accessibility and interdisciplinary collaboration within shared public areas.

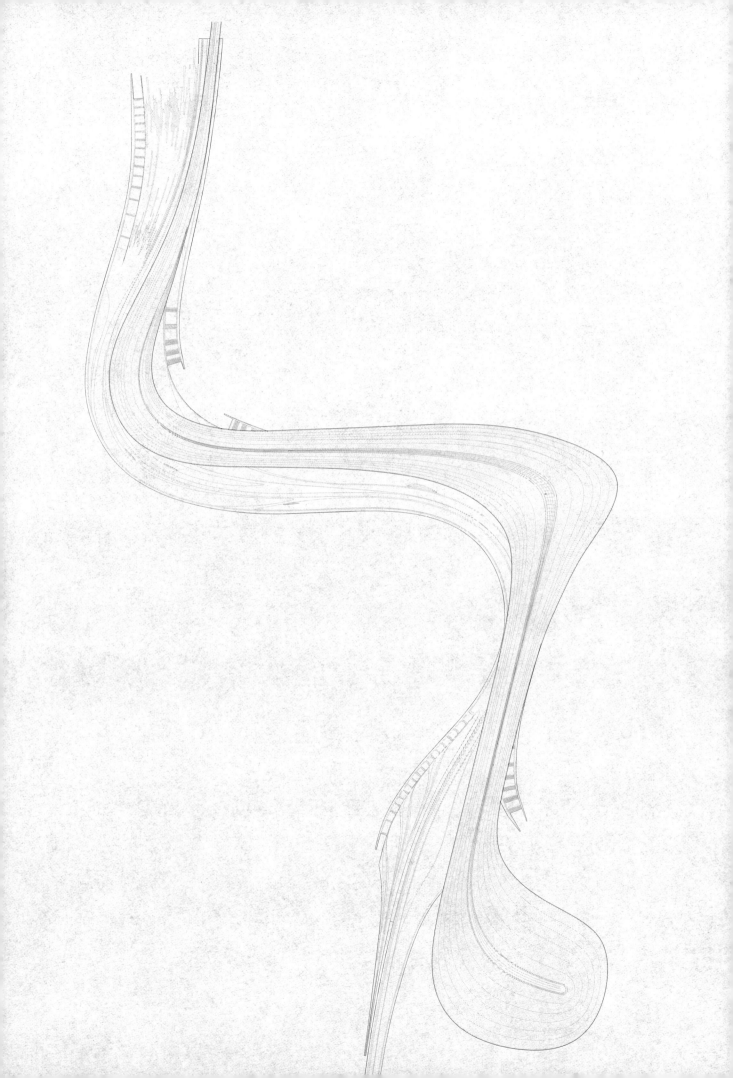

三亚国际艺术港

中国，三亚
2022年（进行中）
三亚中央商务区管理局

三亚国际艺术港位于三亚西南角，毗邻凤凰海岸，展现了这座城市生机勃勃的沿海特征和热带海洋文化。作为三亚文化商务区发展的催化剂，本项目为文化和艺术活动营造了繁荣的发展环境。设计灵感来源于疍家文化。疍家是生活在海上的民族，被称为"海上吉卜赛人"，意在向其坚韧的精神致敬。结合这一背景，规划表达了该城市对沿海区域与自然环境和谐共存的愿景，为项目赋予了特殊的深度和意义。

该综合项目建筑面积为40.9万平方米，由交相呼应的艺术和贸易中心组成，内部设有露天剧场、邮轮码头、商业街和购物广场、商业村与海滨广场，并与沿海景观带联系起来。两座建筑的白色屋顶模仿船帆随风舞动的起伏形态，成为三亚独特的标志。建筑位于场地中心高处，仿佛在宽敞的公共广场上屹立的"双峰"，从海港的任意角度都能牢牢抓住人们的注意力。

该项目秉承绿色、低碳的设计原则，致力于成为一个可持续发展的社区。全部建筑均力求达到绿色三星评级或更高标准，部分区域还树立了超低能耗的典范。依托优越的地理位置，该项目采用了各类清洁能源策略，包括光伏发电、风力发电等。此外，屋顶雨水收集、地下水处理、土壤净化和渗透及地下储存设施等生态处理手法也将为其成为绿色生态建设典范做出贡献。

Sanya International Art Harbour

Sanya, China
2022 (Ongoing)
Sanya Central
Business District Administration

Situated at the southwestern corner of Sanya, near the Phoenix Coast, the Sanya International Art Harbour embodies the city's vibrant coastal identity and its tropical marine culture. As a catalyst for the development of the Sanya Cultural Business District, it fosters a thriving environment for cultural and artistic activities. Drawing inspiration from the Tanka culture, an ethnic group traditionally living on boats by the sea, often known as 'sea gypsies,' the design pays homage to the region's resilient history. Connecting to this context, the design reflects the city's vision of a coastal public space that harmoniously coexists with the environment, adding depth and meaning to the project.

With a total built area of 409,000 square metres, the mixed-use project features an art centre and a trade centre, facing each other. The tapestry of design also incorporates an amphitheatre, a cruise terminal, a retail street and plaza, a commercial village, and a waterfront plaza, establishing a connection with the coastal landscape belt. Emulating the curvature of billowing sails, the white roofs of the two buildings serve as a distinctive symbol for Sanya City. Positioned in an elevated stance at the centre of the site, they form twin peaks above the expansive public plaza, capturing attention from different viewpoints across the harbour.

Embracing green and low-carbon design principles, the project strives to establish a sustainable community. All buildings adhere to Green Three-Star or higher standards, and select units showcase demonstrations of ultra-low energy consumption. Leveraging its advantageous geographical location, the project incorporates clean energy strategies, such as photovoltaics supplemented by wind power. Furthermore, ecological features like roof rainwater collection, ground water treatment, soil purification and infiltration, and buried storage facilities contribute to its status as a demonstration site for green and ecological construction.

夏中心

中国，西安
2023年（进行中）
西安睿兴生置业有限公司

夏中心位于西安市中心西南部的雁塔区，是通往西安经济技术开发区的门户。该塔高210米，模仿潺潺流水的优美律动，呈现出蜿蜒的曲线。夏中心独特的造型与内部宜人的绿色空间相融合，打造出迷人且和谐的环境。

塔楼极具魅力的中庭如瀑布一般，穿插延伸至幕墙。中庭内设有多层次的休憩空间和立体花园，为塔楼增添了一丝优雅与宁静。

夏中心运用先进的技术，在幕墙上集成传感器，根据环境分析、使用位置采用不同反射率的中空玻璃。塔楼外表皮嵌入光伏单元、遮阳百叶和垂直竖框，践行可持续发展的设计。

凭借其引人注目的设计、功能空间和技术创新特点，夏中心将成为西安的一道亮丽风景线，同时也会在周边地区的持续发展中发挥关键作用。

Daxia Tower

Xi'an, China
2023 (Ongoing)
Xi'an Ruixing Sheng Real Estate Co.

Located in Xi'an's Yanta District, southwest of the city centre, the Daxia Tower serves as the gateway to the Xi'an High-tech Economic and Technological Development Zone. Standing tall at 210 metres, the tower displays a sinuous curvature that mimics the graceful flow of cascading water. Its unique profile integrates inviting green spaces within its interior, creating a captivating and balanced environment.

One of the tower's standout features is its captivating atriums, which traverse the façade like a series of waterfalls flowing in succession. These atriums house multilevel breakout spaces and hanging gardens, adding a touch of elegance and tranquillity to the tower's design.

The Daxia Tower showcases advanced façade technologies, incorporating integrated sensors and layered glass surfaces with varying reflectivity and patterns. Its exterior skin seamlessly integrates photovoltaic cells, sun shading louvres, and vertical mullions, harnessing energy and embracing sustainability.

With its striking design, functional spaces, and innovative technological features, the Daxia Tower contributes to the skyline of Xi'an while also playing a pivotal role in the ongoing development of the surrounding region.

元宇宙隧道

⑩

"元宇宙隧道"展示了扎哈·哈迪德建筑事务所(ZHA)在虚拟世界设计的一系列项目,物理世界与数字领域在此交融。这些项目强调了元宇宙的各种可能性:

1. "NFT主义"虚拟画廊,于2021年巴塞尔艺术展迈阿密展会的虚拟展览中推出,主要展出数字艺术品,包括非同质化代币(NFTs)。
2. 利伯兰元宇宙是一个网络城市孵化器和虚拟城市设计,旨在支持克罗地亚与塞尔维亚之间的微型国家(尚未获得国际认可)利伯兰自由共和国的主权发展。它也是Web 3.0产业内的一个网络和合作场所。
3. 建筑元宇宙是沉浸式体验的艺术项目,融合了建筑、艺术、技术和人工智能。该项目由雷菲克·阿纳多尔工作室(RAS)与ZHA合作完成,在2022年首尔东大门设计广场举办的"元视界:未来即现在"展览中展出。该项目采用基于ZHA建筑数据而编写的人工智能算法,从而生成了一个跨越建筑与推测性的空间设计组合。
4. 都城乌托邦是ZHA与ArchAgenda公司在2023年威尼斯建筑双年展虚拟展览上共同推出的元宇宙项目。它致力于成为全球社区的虚拟交流中心,为设计学科之间的互动体验营造逼真的环境。

每个项目都代表着ZHA对元宇宙的探索,展示了想象力、设计与技术的融合,创造了连接人与空间的虚拟沉浸式体验。

Metaverse Tunnel

The Metaverse Tunnel showcases a selection of ZHA's ventures into designing virtual worlds, where the physical and digital realms intersect. The projects featured highlight the diverse possibilities of the metaverse:

1. 'NFTism' is a virtual gallery space that exhibits digital artworks, including non-fungible tokens (NFTs), with a curated virtual exhibition presented at the 2021 Art Basel Miami Beach.
2. The 'Liberland Metaverse' is a cyber-urban incubator and virtual city designed to support the sovereign development of Liberland, a micronation located between Croatia and Serbia. It also serves as a networking and collaboration site within the Web 3.0 industry.
3. 'Architecting the Metaverse' is an immersive art project that combines architecture, art, technology, and artificial intelligence. It was created in collaboration between Refik Anadol Studio (RAS) and Zaha Hadid Architects (ZHA) for the 'Meta-Horizons: The Future Now' exhibition in Seoul in 2022. The project employs AI algorithms trained on ZHA's architectural documentation, resulting in a portfolio of spatial designs spanning the built and the speculative.
4. 'Metrotopia' is a metaverse venture launched by Zaha Hadid Architects and ArchAgenda for the 2023 Venice Architecture Biennale collateral events. It aims to be a virtual communication hub for the global community, providing a lifelike setting for interactive experiences across various design disciplines.

Each project represents ZHA's exploration of the metaverse, showcasing the fusion of imagination, design, and technology in creating immersive virtual experiences that connect people and spaces.

NFT 主义

虚拟画廊
2021年
肯尼·斯坎特

在2021年由 Nagel Draxler 画廊主办的巴塞尔艺术展迈阿密展会上，ZHA 推出了"NFT主义"虚拟画廊，以探索元宇宙中的建筑与社交互动。

赛博空间是一种虚拟环境，可以通过计算机网络实现人与人之间的交流。真实感、三维技术与大型多人在线（MMO）视频游戏技术，结合高速网络与云存储，使赛博空间成为沉浸式、互动式且有社交参与度的三维世界，可以通过多种设备（如桌面浏览器、移动应用程序和智能电视）进行访问。

赛博空间是支撑元宇宙的整体空间网络技术不可或缺的部分——一个将赛博空间的空间交互体验与支持社会、社区形成和经济基础设施相结合的在线环境。元宇宙支持新形式的文化生产和相关体验，例如，数字艺术和虚拟美术馆。

在巴塞尔艺术展迈阿密展会上，虚拟画廊展示了肯尼·沙克特之前委托的设计，其中包括"Z-boat"（限量版快艇）、"Z-Car One"（四座城市汽车）、"Belu"（雕塑长凳）和"Orchis"（凳子）。这些设计均来自扎哈·哈迪德建筑事务所（ZHA）与扎哈·哈迪德设计（ZHD）之间的合作。ZHA 的空间设计侧重于用户体验、社交互动与戏剧性构成，这种虚拟建筑由性能校准、现场测试的参数化设计技术提供支持。ZHA 的设计与 MMO 和交互技术相结合，为广大在线观众带来了全新的用户体验。

NFTism

Virtual Gallery
2021
Kenny Schachter

Presented in 2021 at Art Basel Miami Beach and hosted by the Galerie Nagel Draxler, 'NFTism' is a virtual gallery that explores architecture and social interaction in the metaverse.

Cyberspaces are virtual environments that enable human-to-human communication via computer networks. The current technologies used in photorealistic, 3D, and massively multiplayer online (MMO) video-game creation, combined with high-speed network and cloud technologies, allow cyberspaces to be three-dimensional, interaction-rich, sensorial, socially engaging, and accessible through various devices such as desktop browsers, mobile apps, and smart TVs.

Cyberspaces are integral to the spatial-web technologies that form the foundation of the metaverse, an online environment that combines the spatial and interaction experiences of cyberspaces with supporting social, community-forming, and economic infrastructures. The metaverse enables new forms of cultural production and associated experiences, including digital art and virtual art museums.

Launched at Art Basel Miami Beach in 2021, the virtual gallery showcases, among other items, a collection of designs previously commissioned by Kenny Schachter, namely the 'Z-boat,' the 'Z-Car One,' a sculptural bench-table called 'Belu' and a stool named 'Orchis.' These designs were collaborations between Zaha Hadid Design and Zaha Hadid Architects (ZHA). With spatial designs by ZHA that prioritise user experience, social interaction, and dramaturgical composition, this virtual architecture utilises performance-aligned and field-tested parametric design technologies. By combining ZHA's designs with MMO and interaction technologies, novel and user-enhancing experiences are brought to a wide online audience.

利伯兰元宇宙

Mytaverse
2021年
利伯兰自由共和国

利伯兰自由共和国是一个由维特·耶德利奇卡和亚娜·马尔科维奇乔娃于2015年建立的虚拟微型国家（尚未获得国家认可），设定于多瑙河西岸，是克罗地亚和塞尔维亚之间的一个无主之地，面积约为7平方公里，目前是第三小的"主权国家"，仅次于梵蒂冈和摩纳哥。

利伯兰在2021年首次通过虚拟元宇宙向其电子公民和世界各地的支持者正式公布，并计划未来在物理世界中实现。利伯兰元宇宙与其他元宇宙的区别在于其更专注于加密货币与区块链技术生态系统，而非娱乐行业。投资利伯兰元宇宙将获得利伯兰实体的股份。

利伯兰元宇宙的城市设计和建筑设计也十分突出，它打造了一个互动和沉浸式的三维空间环境。城市结构以宽阔的开放空间和活跃的户外公共区域为特色，这些区域从中央商务区辐射而来，由一系列活动场所组成，如NFT广场、DeFi广场、展览中心、孵化园与市政厅。城市治理模式适用于利伯兰的虚拟与物理领域，凭借多样的规划原则为开发商、投资者、买家和终端用户提供选择。利用各种收入和所有权结构，利伯兰在不同领域对这些原则进行探索，包括赞助秩序、自治秩序和自发秩序外围。

利伯兰代表了将元宇宙和Web 3.0愿景与城市规划相结合的一种尝试，采用了一种去中心化和开放式的参与模式，将空间和治理技术相融合。它展示了虚拟建筑、城市化和现实生活体验之间的相互关系，强调了这一技术如何丰富公民的体验和促进经济的繁荣。

Mytaverse作为一个企业级3D沉浸式平台，为该项目的沉浸式虚拟现实互动和通信技术提供支持。ZHA的设计在其中适得其所，并使利伯兰焕发活力。

Liberland Metaverse

Mytaverse
2021
Liberland

The Free Republic of Liberland (Liberland) is a micro-nation located on the West Bank of the Danube River, in an unclaimed territory between Croatia and Serbia. Founded in 2015 by Vít Jedlička and Jana Markovičová, Liberland is approximately 7 square kilometres and is now the third smallest 'sovereign' state, after the Vatican and Monaco.

The nation first launched to its e-citizens and supporters around the world in 2021 through a virtual metaverse with a physical launch planned for the future. Liberland Metaverse differentiates itself from contemporary metaverses by focusing on the crypto and the blockchain technology ecosystem rather than the entertainment sector. Investing in Liberland Metaverse grants a stake in the physical Liberland.

Similarly, Liberland Metaverse stands out for its urban and architectural design, creating an interactive and immersive 3D spatial environment. The urban fabric features broad open spaces and activated outdoor public areas that radiate from a Central Business District, consisting of a series of event venues such as NFT plaza, DeFi plaza, Exhibition Centre, Incubator, and City Hall. The urban governance model applies to both virtual and physical realms of Liberland, leveraging a variety of planning principles to provide options for developers, investors, buyers, and end users. Using a variety of revenue and ownership structures, these principles are explored in different districts of Liberland, including sponsored order, self-governed order, and spontaneous order outskirts.

Liberland represents an attempt to merge the metaverse and web 3.0 vision with urban planning, adopting a decentralised and open participatory model that integrates spatial and governance technologies. It demonstrates the mutual relationship between virtual architecture, urbanism, and in real life (IRL) experiences, highlighting how this integration can enrich and enhance the citizenry's experience and economic prosperity.

The Liberland Metaverse venture utilises Mytaverse, which powers the immersive VR interaction and communication technology. The designs of Zaha Hadid Architects are deployed within Mytaverse, an enterprise 3D immersive platform employed by Liberland to bring the city to life.

建筑元宇宙

2022年
首尔设计基金会

"建筑元宇宙"是沉浸式体验的艺术项目，融合了建筑、艺术、技术和人工智能。该项目是为2022年首尔东大门设计广场举办的"元视界：未来即现在"展览创作的，是雷菲克·阿纳多尔工作室(RAS)和ZHA为期六个月合作的成果。

通过这一独特的项目，RAS与ZHA创造了一件融合复杂技术和崇高美学的艺术作品，同时连接了网络物理架构与人工智能。"建筑元宇宙"扩展了RAS媒体实验室正在进行的研究项目，并在ZHA的作品中将整个建筑数据库文献可视化。"多模型"人工智能算法被用于分析ZHA编译和合成的图像与脚本数据。通过算法得出的综合数据集象征着跨越实体建筑与虚拟空间的设计组合。

该项目是ZHA与RAS的首次合作，两家公司在各自的建筑、艺术领域都是代表最前沿技术的实践先驱。"建筑元宇宙"是第一个让ZHA建筑作品中的"机器梦"遍布全球的项目，通过ZHA备受推崇的设计与人工智能结合创造出的沉浸式空间体验呈现。

Architecting the Metaverse

2022
Seoul Design Foundation

Architecting the Metaverse is an immersive art project at the intersection of architecture, art, technology, and artificial intelligence (AI). Created on occasion of the 2022 'Meta-Horizons: The Future Now' exhibition at DDP in Seoul, it is the result of a 6-month collaboration between Refik Anadol Studio (RAS) and Zaha Hadid Architects (ZHA).

With this unique project, RAS and ZHA create an artwork that merges complexity with sublime aesthetics while simultaneously connecting various discourses in cyber-physical architecture and artificial intelligence. Architecting the Metaverse extends RAS Media Lab's ongoing research project and visualises their entire database of architectural documentation in the oeuvre of ZHA. 'Multi-model' AI algorithms have been utilised to analyse image and script data compiled and composed by ZHA. These algorithms result in comprehensive datasets that represent a portfolio of spatial designs spanning the built and the speculative.

This project marks the first collaboration between RAS and ZHA—each a pioneer in their respective fields of architecture, art, and technology. Architecting the Metaverse is the first of its kind in making the machine dream about ZHA architectural works around the world, presented through an immersive spatial experience generated by ZHA's venerated designs and AI.

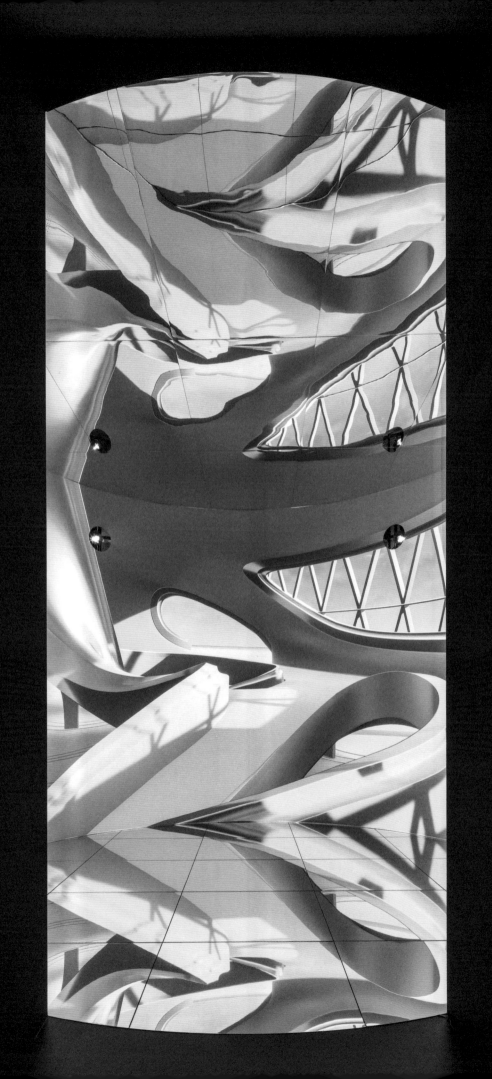

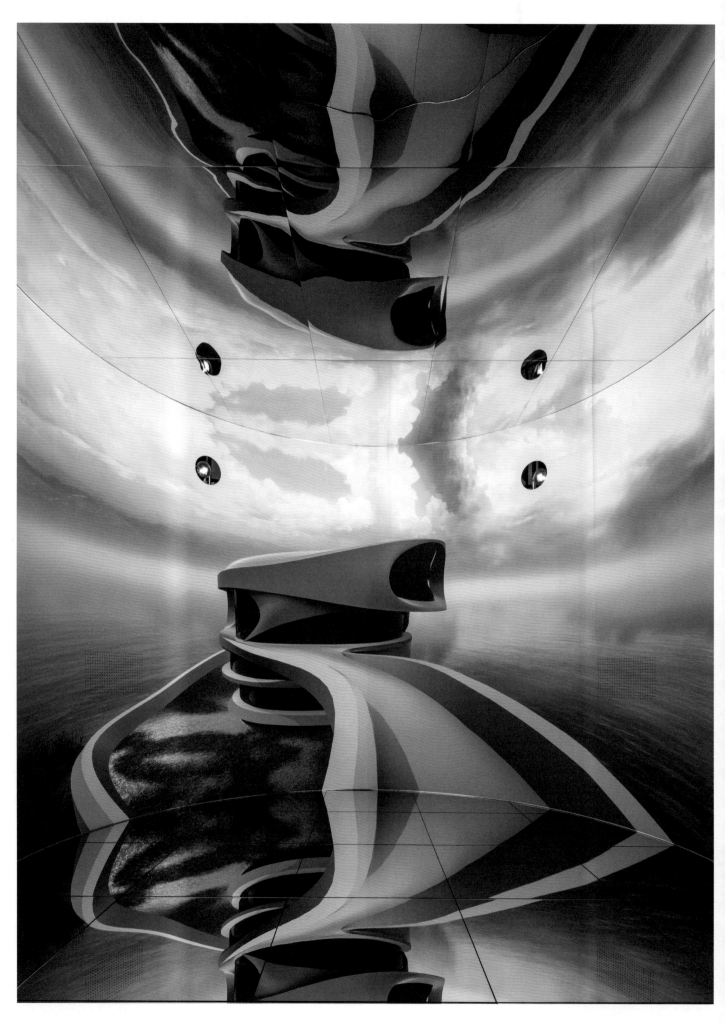

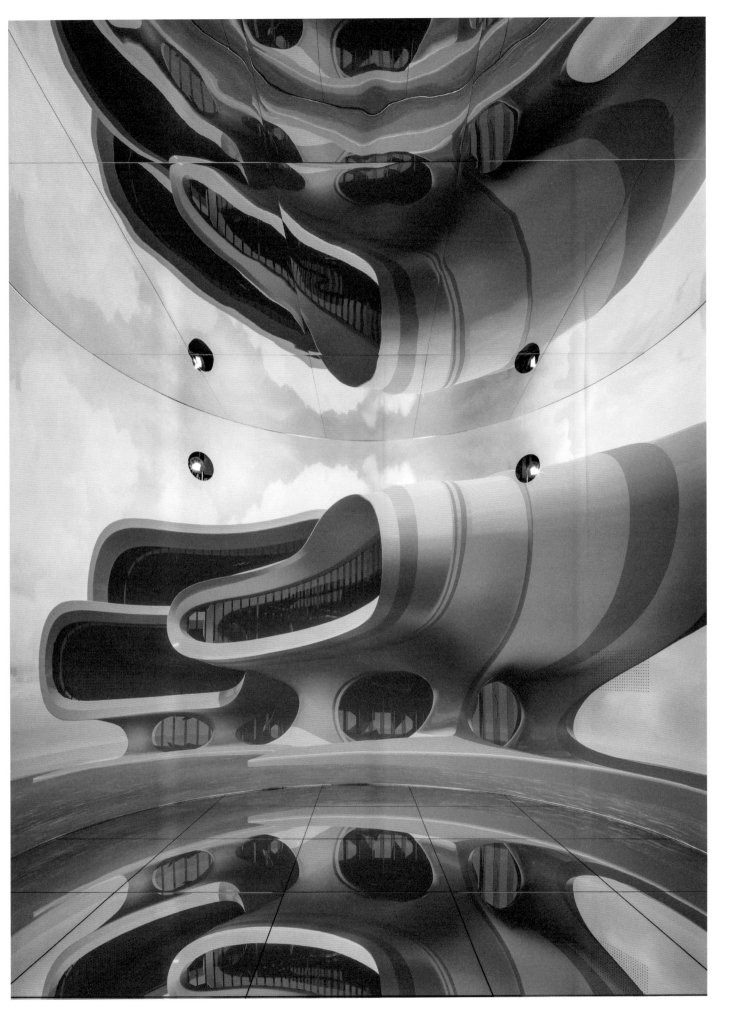

都城乌托邦

2023年
Metrotopia

ZHA与总部位于芝加哥的ArchAgenda公司共同开发的"都城乌托邦"是一个独特的元宇宙项目。都城乌托邦致力于成为全球设计界的虚拟交流中心。

都城乌托邦于威尼斯双年展虚拟展览"知识转移"首次亮相。这次展览呈现了全球领先的建筑工作室的近期作品，包括大都会建筑事务所、墨菲西斯建筑事务所、蓝天组建筑事务所、UnStudio建筑事务所、MAD建筑事务所、藤本壮介建筑设计事务所、上海创盟国际建筑设计有限公司、LAVA建筑设计事务所、CAP建筑事务所、EcoLogic Studio建筑和设计工作室，以及扎哈·哈迪德建筑事务所。

该展览回应了由莱斯利·洛科为2023年威尼斯建筑双年展提出的主题"未来的实验室"。都城乌托邦是建筑双年展同期活动之一，其他活动还包括CityX——一个位于威尼斯云端的虚拟展馆群，旨在展示全球建筑学教师的研究工作，而由纽约理工学院主办的在Palazzo Ça Zenobio的实体展览则展示各建筑学院的学生作品。都城乌托邦还将参加2023年芝加哥建筑双年展。

都城乌托邦是使用虚幻引擎5构建的多人虚拟环境，可在逼真的环境中呈现真实社交活动的空间音效，并能让设计实体与空间的特质精准可视化。都城乌托邦的技术合作伙伴与托管平台是Mytaverse。

都城乌托邦力图成为一座虚拟都城，成为创意产业、文化活动与设计生态的中心。它会集了所有设计领域，包括城市、建筑、室内、家具、产品、时装、平面设计，还包括相关的学校、画廊、博物馆、展览、设计周、双年展、奖项和会议，以及相关的出版社、杂志与网络媒体。

Metrotopia

2023
Metrotopia

Metrotopia is a unique metaverse venture, dedicated to becoming the virtual communication hub for the global community, launched by Zaha Hadid Architects and Chicago-based ArchAgenda.

The Metrotopia Metaverse debuted with a virtual exhibition entitled 'Knowledge Transfer' at the Venice Architecture Biennale. The exhibition showcased the works of renowned architecture studios such as OMA, Morphosis, Coop Himmelb(l)au, UnStudio, MAD, Sou Fujimoto, Archi-Union, LAVA, CAP, EcoLogic Studio and ZHA.

The exhibition aligns with the theme of the 2023 Venice Architecture Biennale, 'The Laboratory of the Future.' Metrotopia is a part of the Architecture Biennale collateral events that include CityX and a physical exhibition at Palazzo Ça Zenobio. CityX is a swarm of virtual pavilions hovering above Venice that display the research work of architecture teachers worldwide, while the exhibition at Palazzo Ça Zenobio, hosted by the New York Institute of Technology, showcases student works from various schools of architecture. Metrotopia will also contribute to the Chicago Architecture Biennial later this year.

Metrotopia Metaverse is a multi-player virtual environment built with Unreal Engine 5. It features spatial sound for realistic communicative interaction within a lifelike setting, which allows the character and quality of designed objects and spaces to be faithfully visualised. Metrotopia's technology partner and hosting platform is Mytaverse.

Aspiring to become a virtual city, Metrotopia is a central hub for creative industries, cultural activities, and the wider design ecosystem. It brings together all design disciplines including urban design, architecture, interiors, furniture design, product design, fashion design, graphic design, plus related schools, galleries, museums, exhibitions, design weeks, biennales, awards and conferences, as well as related publishing houses, magazines and web-media.

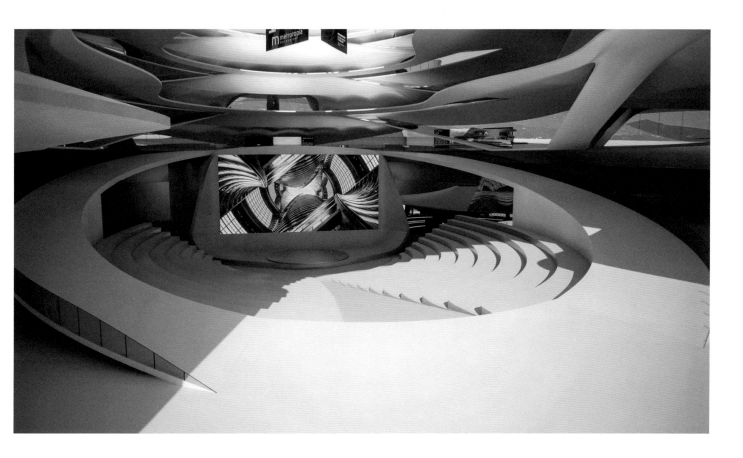

扎哈·哈迪德建筑事务所经典作品年表

Zaha Hadid Architects Timeline

⑪

1990

季风餐厅
日本，札幌
Moonsoon Restaurant
Sapporo, Japan

1993

维特拉消防站
德国，莱茵河畔魏尔
Vitra Fire Station
Weil am Rhein, Germany

1999

国家园艺展馆
德国，莱茵河畔魏尔
Land Formation One
Weil am Rhein, Germany

2001

霍恩海姆-诺德多联式交通枢纽
法国，斯特拉斯堡
Hoenheim-Nord Terminus and Car Park
Strasbourg, France

2003

伯吉瑟尔滑雪跳台
奥地利，因斯布鲁克
Bergisel Ski Jump
Innsbruck, Austria

2005

罗森塔尔当代艺术中心
美国，辛辛那提
Rosenthal Center for
Contemporary Art
Cincinnatti, USA

宝马汽车总部
德国，莱比锡
BMW Central Building
Leipzig, Germany

费诺科学中心
德国，沃尔夫斯堡
Phaeno Science Centre
Wolfsburg, Germany

奥德鲁普加德博物馆扩建
丹麦，哥本哈根
Ordrupgaard Museum Extension
Copenhagen, Denmark

2007

北山公园缆车站
奥地利，因斯布鲁克
Nordpark Cable Railway Stations
Innsbruck, Austria

2008

香奈儿当代移动艺术馆
中国，香港；日本，东京；美国，纽约
Mobile Art—Chanel
Contemporary Art Container
Hongkong, China; Tokyo, Japan; New York, USA

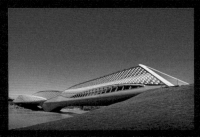

萨拉戈萨廊桥馆
西班牙，萨拉戈萨
Zaragoza Bridge Pavilion
Zaragoza, Spain

2009

MAXXI：国立二十一世纪美术馆
意大利，罗马
MAXXI: Museum of XXI Century Arts
Rome, Italy

2010

伊芙琳格蕾丝学院
英国，伦敦
Evelyn Grace Academy
London, UK

谢赫扎耶德大桥
阿联酋，阿布扎比
Sheikh Zayed Bridge
Abu Dhabi, UAE

广州大剧院
中国，广州
Guangzhou Opera House
Guangzhou, China

2011

格拉斯哥交通博物馆
英国，格拉斯哥
Glasgow Riverside Museum of Transport
Glasgow, UK

达飞海运集团总部
法国，马赛
CMA CGM Headquarters
Marseille, France

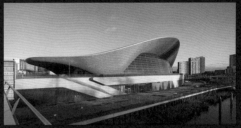

伦敦水上运动中心
英国，伦敦
London Aquatics Centre
London, UK

2012

皮埃尔·韦斯大厦
法国，蒙彼利埃
Pierresvives
Montpellier, France

伊莱和伊迪特·布罗德美术馆
美国，东兰辛
Eli and Edythe Broad Art Museum
East Lansing, USA

银河SOHO
中国，北京
Galaxy SOHO
Beijing, China

2013

阿利耶夫中心
阿塞拜疆，巴库
Heydar Aliyev Centre
Baku, Azerbaijan

蛇形北画廊
英国，伦敦
Serpentine North Gallery
London, UK

维也纳经济大学图书馆与学习中心
奥地利，维也纳
Library and Learning Centre, University of Economics Vienna
Vienna, Austria

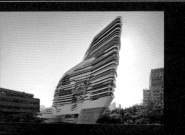

2014

赛马会创新楼
中国，香港
**Jockey Club Innovation Tower,
Hong Kong Polytechnic University**
Hong Kong, China

东大门设计广场
韩国，首尔
Dongdaemun Design Plaza
Seoul, South Korea

望京 SOHO
中国，北京
Wangjing SOHO
Beijing, China

2015

丽敦豪邸
新加坡，新加坡市
d'Leedon
Singapore, Singapore

凌空 SOHO
中国，上海
Sky SOHO
Shanghai, China

牛津大学圣安东尼学院中东中心
英国，牛津
**Investcorp Building for
Oxford University at
St Antony's College**
Oxford, UK

梅斯纳尔山科罗内斯博物馆
意大利，南蒂罗尔
Messner Mountain Museum Corones
South Tyrol, Italy

2016 / 2017

自治领大厦
俄罗斯，莫斯科
Dominion Office Building
Moscow, Russia

萨勒诺海运码头
意大利，萨勒诺
Salerno Maritime Terminal
Salerno, Italy

安特卫普港屋顶大楼
布鲁塞尔，安特卫普
Port House
Antwerp, Belgium

那不勒斯 - 阿夫拉戈拉新高铁站
意大利，那不勒斯
Napoli-Afragola High Speed Train Station
Naples, Italy

2018

阿卜杜拉国王石油研究中心
沙特阿拉伯，利雅得
King Abdullah Petroleum Studies and Research Centre
Riyadh, Saudi Arabia

纽约公寓
美国，纽约
520 West 28th
New York City, USA

忠利集团大厦
意大利，米兰
Generali Tower
Milan, Italy

沐梵世度假酒店
中国，澳门
**Morpheus Hotel and
Resort at City of Dreams**

2019

南京国际青年文化中心
中国，南京
Nanjing International Youth Cultural Centre
Nanjing, China

长沙梅溪湖国际文化艺术中心
中国，长沙
Changsha Meixihu International Culture and Arts Centre
Changsha, China

丽泽SOHO
中国，北京
Leeza SOHO
Beijing, China

大兴国际机场
中国，北京
Beijing Daxing International Airport
Beijing, China

2020

奥普斯大厦
阿联酋，迪拜
Opus
Dubai, UAE

千号馆
美国，迈阿密海滩
One Thousand Museum
Miami Beach, USA

2021

纬壹科技城
新加坡，新加坡市
One North Masterplan
Singapore, Singapore

天空花园
斯洛伐克共和国，布拉迪斯拉发
Culenova Skypark
Bratislava, Slovakia

广州无限极广场
中国，广州
Guangzhou Infinitus Plaza
Guangzhou, China

埃莱夫塞里亚广场
塞浦路斯，尼科西亚
Eleftheria Square
Nicosia, Cyprus

2022

碧哈总部
阿联酋，沙迦
BEEAH Headquarters
Sharjah, UAE

2023

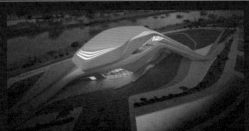

拉巴特大剧院
摩洛哥，拉巴特
Grand Théâtre de Rabat
Rabat, Morocco

西一线跨绛溪河大桥
中国，成都
Chengdu West Line Cross-Jiangxi River Bridge
Chengdu, China

西安国际足球中心
中国，西安
Xi'an International Football Centre
Xi'an, China

澳门新濠影汇二期
中国，澳门
Studio City Phase 2
Macau, China

成都科幻馆
中国，成都
Chengdu Science Fiction Museum
Chengdu, China

珠海金湾艺术中心
中国，珠海
Zhuhai Jinwan Civic Art Centre
Zhuhai, China

图片版权

摄影由以下人员/团队提供

Aaron Peacock© 115
Arch-Exist Photography© 272-273, 291, 292-293, 375, 376
Christian Richters/VIEW© 89, 90-91, 194
Cat-Optogram© 302-303, 376
Fernando Guerra© 199, 372
Hanne Hvattum© 163
Hawkaye© 195
Hélène Binet© 99, 199, 203, 372, 372, 372, 372, 373, 374, 374
Hufton + Crow© 103, 104-105, 107, 108-109, 117, 118-119, 121, 123, 124, 125, 127, 128-129, 135, 136-137, 139, 140-141, 142-143, 186, 187, 190, 191, 194, 198, 199, 201, 203, 206, 206, 206, 229, 230-231, 239, 240-241, 251, 252, 253, 254, 255, 372, 373, 373, 373, 373, 373, 373, 374, 374, 374, 374, 374, 374, 375, 375, 375, 375
IG: Seven 7 Panda© 234
Illulian© 149
Iwan Baan© 100-101, 215, 216-217, 219
Jerry Yin/SOHO China© 374
John Linden© 97, 372
Ju Huanzong/Xinhua/Alamy Live News© 247, 248-249, 375
Kyungsub Shin© 365, 366, 367
Laurian Ghinitoiu© 131, 132-133, 375, 375
Luke Hayes© 155, 155, 157, 189, 372, 373, 373, 374
Matt Danby© 163
Matt Walker, Youness Yousefi, Jo-Lynn Yen, Anagha Patil, Alicia Nahmad© 161, 165
McAteer Photograph/Alan McAteer© 195, 373
Paul Warchol© 85, 86, 87, 372, 373
Penta Investments© 375
Roland Halbe© 372, 372, 372
Studio Naaro© 153
Virgile Simon Bertrand© 46-47, 52-53, 54-55, 56-57, 58-59, 60-61, 66-67, 70-71, 73, 74-75, 76, 111, 112-113, 220, 221, 223, 224-225, 226-227, 233, 235, 243, 244-245, 263, 264-265, 374, 374, 374, 375
Werner Huthmacher© 93, 94-95, 199, 372, 372, 374
Xue Liang© 267, 268, 269, 271, 375
张灏 Seilaojiong© 301
Zhu Yumeng© 48-49, 50-51, 62-63, 64-65, 68, 69, 72, 77, 169, 210-211

效果图及第三方可视化数据由以下人员/团队提供

Atchain: 189, 258, 276-277, 291, 292-293, 327, 328, 329, 335, 336-337, 347, 348-349, 355, 356-357
Brick Visual: 259, 283, 284-285, 287, 339, 340-341, 376, 376
Clover: 319, 321
Cosmocube: 311, 312-313
Frontop: 306-307
Metrotopia Metaverse: 367, 367
Minmud: 280-281, 302-303, 376
MIR: 203, 279, 305, 309
Mytaverse: 363
Negativ: 145, 146-147, 203, 295, 297, 298, 315, 316-317, 331, 332-333, 352-353
OmegaRender: 155
Plomp: 344-345, 351
Proloog: 299, 299, 343
pyxid: 323, 324-325
Slaschcube: 185, 185, 185, 257, 288, 289
www.openstreetmap.org/copyright: 170
www.researchgate.net/figure/Census-geographic-unit-population-densities-of-Beijing-urban-area_fig6_236666871: 170

除以上说明，所有图片均由扎哈·哈迪德建筑事务所提供

Image Credits

Photographs are courtesy of

Aaron Peacock© 115
Arch-Exist Photography© 272-273, 291, 292-293, 375, 376
Christian Richters/VIEW© 89, 90-91, 194
Cat-Optogram© 302-303, 376
Fernando Guerra© 199, 372
Hanne Hvattum© 163
Hawkaye© 195
Hélène Binet© 99, 199, 203, 372, 372, 372, 372, 373, 374, 374
Hufton + Crow© 103, 104-105, 107, 108-109, 117, 118-119, 121, 123, 124, 125, 127, 128-129, 135, 136-137, 139, 140-141, 142-143, 186, 187, 190, 191, 194, 198, 199, 201, 203, 206, 206, 206, 229, 230-231, 239, 240-241, 251, 252, 253, 254, 255, 372, 373, 373, 373, 373, 373, 373, 374, 374, 374, 374, 374, 374, 375, 375, 375, 375
IG: Seven 7 Panda© 234
Illulian© 149
Iwan Baan© 100-101, 215, 216-217, 219
Jerry Yin/SOHO China© 374
John Linden© 97, 372
Ju Huanzong/Xinhua/Alamy Live News© 247, 248-249, 375
Kyungsub Shin© 365, 366, 367
Laurian Ghinitoiu© 131, 132-133, 375, 375
Luke Hayes© 155, 155, 157, 189, 372, 373, 373, 374
Matt Danby© 163
Matt Walker, Youness Yousefi, Jo-Lynn Yen, Anagha Patil, Alicia Nahmad© 161, 165
McAteer Photograph/Alan McAteer© 195, 373
Paul Warchol© 85, 86, 87, 372, 373
Penta Investments© 375
Roland Halbe© 372, 372, 372
Seilao Jiong© 301
Studio Naaro© 153
Virgile Simon Bertrand© 46-47, 52-53, 54-55, 56-57, 58-59, 60-61, 66-67, 70-71, 73, 74-75, 76, 111, 112-113, 220, 221, 223, 224-225, 226-227, 233, 235, 243, 244-245, 263, 264-265, 374, 374, 374, 375
Werner Huthmacher© 93, 94-95, 199, 372, 372, 374
Xue Liang© 267, 268, 269, 271, 375
Zhu Yumeng© 48-49, 50-51, 62-63, 64-65, 68, 69, 72, 77, 169, 210-211

Renders and research are courtesy of

Atchain: 189, 258, 276-277, 291, 292-293, 327, 328, 329, 335, 336-337, 347, 348-349, 355, 356-357
Brick Visual: 259, 283, 284-285, 287, 339, 340-341, 376, 376
Clover: 319, 321
Cosmocube: 311, 312-313
Frontop: 306-307
Metrotopia Metaverse: 367, 367
Minmud: 280-281, 302-303, 376
MIR: 203, 279, 305, 309
Mytaverse: 363
Negativ: 145, 146-147, 203, 295, 297, 298, 315, 316-317, 331, 332-333, 352-353
OmegaRender: 155
Plomp: 344-345, 351
Proloog: 299, 299, 343
pyxid: 323, 324-325
Slaschcube: 185, 185, 185, 257, 288, 289
www.openstreetmap.org/copyright: 170
www.researchgate.net/figure/Census-geographic-unit-population-densities-of-Beijing-urban-area_fig6_236666871: 170

Unless otherwise stated, all images are courtesy and copyright of Zaha Hadid Architects

团队名单

扎哈·哈迪德建筑事务所（设计与策展团队）
总裁：Patrik Schumacher
项目总监：Satoshi Ohashi, Manon Janssens, Yang Jingwen
执行负责人：Ziyan Xu, Nan Jiang, Xuexin Duan
项目团队：Sanxing Zhao, Mingjia Zhang, Du Huang, Shuaiwei Li, Qiyue Li, Shuchen Dong, Pittayapa Suriyapee, Ignacio Garcia Martinez, Juan Liu, Yuling Ma, Pengcheng Gu, Daria Zolotareva, Congyue Wang, Feifei Fan, Yihui Wu, Li li, Haotian Man, Felix Amiss
视频：Henry Virgin
算法设计：Taizhong Chen, Yutong Xia, Ling Mao, Taeyoon Kim, Jennifer Durand, Keerti Manney, Jianfei Chu, Vishu Bhooshan, Henry Louth, Jose Pareja Gomez, Shajay Bhooshan
数字社会：Tyson Hosmer, Baris Erdincer, Ziming He
可持续：
• Carlos Bausa Martinez, Aditya Ambare, Aleksander Mastalski, Bahaa Alnassrallah, Disha Shetty, Shibani Choudhury
• Marie-Perrine Plaçais, Melodie Leung, Nica Sabet
编辑与翻译：Ziyan Xu, Amelie Yizhou Liu, Daria Zolotareva, Congyue Wang, Juan Liu, Nan Jiang, Xuexin Duan, Rui Li
海报设计：Patrik Schumacher, Melodie Leung, Nastasja Mitrovic, Daniela Bedoya, Shuaiwei Li, Ceren Tekin, Nica Sabet, Alex Turner
媒体与宣传：Roger Howie, Emma Hawes
图录：Manon Janssens, Daria Zolotareva, Amelie Yizhou Liu, Ziyan Xu, Congyue Wang, Henry Virgin, Juan Liu, Nan Jiang, Xuexin Duan, Rui Li
所有ZHA项目的名单请查看 www.zaha-hadid.com

嘉德艺术中心
展览总策划：寇勤、王彤
展览总监：夏璐
展览团队：李光利、李晴芳、李鹏、任荷
出版团队：杨涓、杨慕涵、严冰
展览协调：陈树银、侯晋晋、郭振兴、马学东、蒲宇、张磊、薛炜、王卓然、肖雅馨、宋海涛、张志欣、杨玲、景松、朱狄帆、张东涛、王思佳、马鑫悦、张爽、姜月、王颖、张颖、水梓辰
展览图录出版：广西师范大学出版社
图录设计：M^{oo} Design

特别鸣谢
数字战略合作伙伴：丝路视觉科技股份有限公司
行业深度合作伙伴：PPG建筑涂料
行业特别合作伙伴：爽物业有限公司 / 高仪SPA
技术合作伙伴：Epson / 京东方 / 中联超清 / 大界
合作伙伴：携程旅行 / 北玻股份 / 璞瑄酒店 / ILLULIAN / kvadrat / 熙玛 / 沈阳远大 / 博丽科技 / 北京建筑大学 / 博威特 / 丰美传媒
首席艺术合作媒体：玩家惠 ARTnews
独家内容合作平台：小红书
特别合作媒体：卷宗 Wallpaper

Exhibition Credits

Zaha Hadid Architects (Design and Curation)
Principal: Patrik Schumacher
Project Directors: Satoshi Ohashi, Manon Janssens, Yang Jingwen
Exhibition Leads: Ziyan Xu, Nan Jiang, Xuexin Duan
Project Team: Sanxing Zhao, Mingjia Zhang, Du Huang, Shuaiwei Li, Qiyue Li, Shuchen Dong, Pittayapa Suriyapee, Ignacio Garcia Martinez, Juan Liu, Yuling Ma, Pengcheng Gu, Daria Zolotareva, Congyue Wang, Feifei Fan, Yihui Wu, Li li, Haotian Man, Felix Amiss
Videos: Henry Virgin
ZHA CODE: Taizhong Chen, Yutong Xia, Ling Mao, Taeyoon Kim, Jennifer Durand, Keerti Manney, Jianfei Chu, Vishu Bhooshan, Henry Louth, Jose Pareja Gomez, Shajay Bhooshan
ZHA Social: Tyson Hosmer, Baris Erdincer, Ziming He
Sustainability:
• Carlos Bausa Martinez, Aditya Ambare, Aleksander Mastalski, Bahaa Alnassrallah, Disha Shetty, Shibani Choudhury
• Marie-Perrine Plaçais, Melodie Leung, Nica Sabet
Editing and translations: Ziyan Xu, Amelie Yizhou Liu, Daria Zolotareva, Congyue Wang, Juan Liu, Nan Jiang, Xuexin Duan, Rui Li
Poster Design: Patrik Schumacher, Melodie Leung, Nastasja Mitrovic, Daniela Bedoya, Shuaiwei Li, Ceren Tekin, Nica Sabet, Alex Turner
Press: Roger Howie, Emma Hawes
Catalogue: Manon Janssens, Daria Zolotareva, Amelie Yizhou Liu, Ziyan Xu, Congyue Wang, Henry Virgin, Juan Liu, Nan Jiang, Xuexin Duan, Rui Li
All ZHA project credits can be found at www.zaha-hadid.com

Guardian Art Center
Chief Exhibition Curators: 寇勤、王彤
Exhibition Director: 夏璐
Exhibition Team: 李光利、李晴芳、李鹏、任荷
Publishing Team: 杨涓、杨慕涵、严冰
Exhibition Coordination: 陈树银、侯晋晋、郭振兴、马学东、蒲宇、张磊、薛炜、王卓然、肖雅馨、宋海涛、张志欣、杨玲、景松、朱狄帆、张东涛、王思佳、马鑫悦、张爽、姜月、王颖、张颖、水梓辰
Exhibition Catalogue published by: Guangxi Normal University Press
Catalogue Design: M^{oo} Design

Special Thanks
Digital Strategy Partner: 丝路视觉科技股份有限公司
In-depth industry partners: PPG建筑涂料
Industry Special Partners: 爽物业有限公司 / 高仪SPA
Technology partners: Epson / 京东方 / 中联超清 / 大界
Partners: 携程旅行 / 北玻股份 / 璞瑄酒店 / ILLULIAN / kvadrat / 熙玛 / 沈阳远大 / 博丽科技 / 北京建筑大学 / 博威特 / 丰美传媒
Chief Art Cooperation Media: 玩家惠 ARTnews
Exclusive content cooperation platform: 小红书
Special partner media: 卷宗 Wallpaper

图书在版编目(CIP)数据

未来之境：扎哈·哈迪德建筑事务所设计展 / 英国扎哈·哈迪德建筑事务所，嘉德艺术中心编 .—桂林：广西师范大学出版社，2024.4

ISBN 978-7-5598-6603-5

Ⅰ．①未… Ⅱ．①英… ②嘉… Ⅲ．①建筑设计－作品集－英国－现代 Ⅳ．① TU206

中国国家版本馆 CIP 数据核字 (2023) 第 222295 号

未来之境：扎哈·哈迪德建筑事务所设计展
WEILAI ZHI JING: ZHAHA·HADIDE JIANZHU SHIWUSUO SHEJIZHAN

出 品 人：刘广汉
策　　划：寇 勤　李 昕
特约编辑：杨 涓　杨慕涵　严 冰
责任编辑：冯晓旭
特约审校：李 倩　张雪梅
装帧设计：刘 伟

广西师范大学出版社出版发行

（广西桂林市五里店路 9 号　邮政编码：541004）
（网址：http://www.bbtpress.com）

出版人：黄轩庄
全国新华书店经销
销售热线：021-65200318　021-31260822-898
北京雅昌艺术印刷有限公司印刷
（北京市顺义区高丽营镇金马园达盛路 3 号 邮政编号：101300）
开本：889 mm×1 194 mm　　1/16
印张：27.375　　插页：10　　字数：420 千
2024 年 4 月第 1 版　　2024 年 4 月第 1 次印刷
定价：488.00 元

如发现印装质量问题，影响阅读，请与出版社发行部门联系调换。